FreeCAD

Basics Tutorial

Tutorial Books

Download Resource Files from:

www.tutorialbook.info

Contents

INTRODUCTION

FreeCAD as a topic of learning is very vast, and having a wide scope. It is a package of many workbenches delivering a great value to enterprises. It offers a set of tools, which are easy-to-use to design, document and simulate 3D models. Using this software, you can design your products at free of cost.

This book provides a step-by-step approach for users to learn FreeCAD. It is aimed for those with no previous experience with FreeCAD. The user will be guided from starting a FreeCAD session to creating parts, assemblies, and drawings. Each chapter has components explained with the help of real-world models.

Scope of this book

This book is written for students and engineers who are interested to learn FreeCAD for designing mechanical components and assemblies, and then create drawings.

This book provides a step-by-step approach for learning FreeCAD. The topics include Getting Started with FreeCAD, Basic Part Modeling, Creating Assemblies, Additional Modeling Tools, and Creating Drawings.

Chapter 1 introduces FreeCAD. The user interface and terminology are discussed in this chapter.

Chapter 2 takes you through the creation of your first FreeCAD model. You create simple parts.

Chapter 3 teaches you to create assemblies. It explains the Top-down and Bottom-up approaches for designing an assembly. You create an assembly using the Bottom-up approach.

Chapter 4: In this chapter, you will learn the sketching tools.

Chapter 5: In this chapter, you will learn additional modeling tools to create complex models.

Chapter 6 teaches you to create drawings of the models created in the earlier chapters.

Chapter 1: Getting Started with FreeCAD

This tutorial book brings in the most commonly used features of FreeCAD.

In this chapter, you will:

- Understand the FreeCAD terminology
- Start a new file
- Understand the User Interface
- Understand different workbenches in FreeCAD

In FreeCAD, you create 3D parts and use them to create 2D drawings and 3D assemblies.

FreeCAD is Feature Based. Features are shapes that are combined to build a part. You can modify these shapes individually.

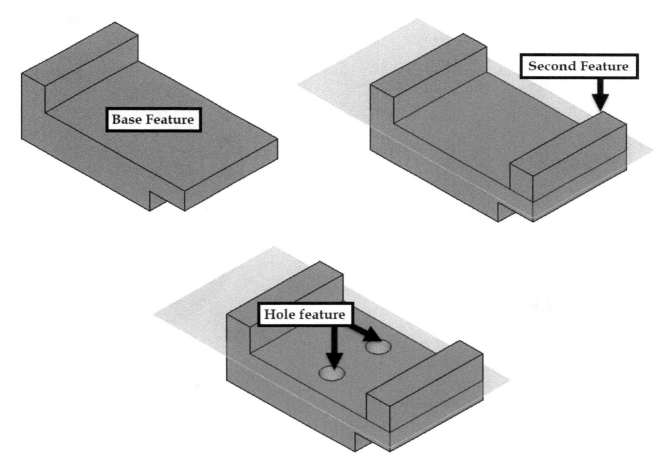

Most of the features are sketch-based. A sketch is a 2D profile and can be extruded, revolved, or swept along a path to create features.

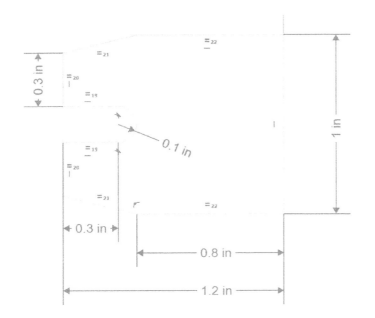

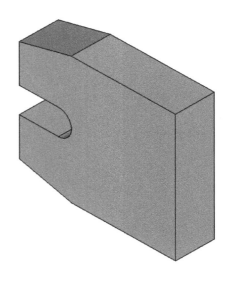

FreeCAD is parametric in nature. You can specify standard parameters between the elements. Changing these parameters changes the size and shape of the part. For example, see the design of the body of a flange before and after modifying the parameters of its features.

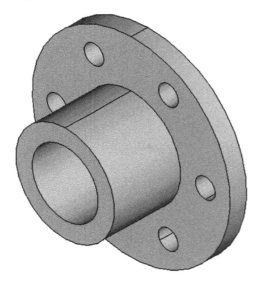

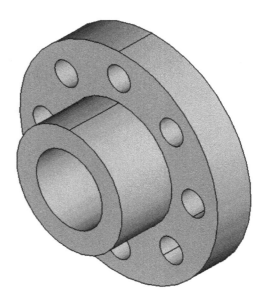

Starting FreeCAD

- Click the Windows icon on the taskbar.
- Click **F > FreeCAD 0.17 > FreeCAD**.
- On the ribbon, click **File > New** to start a new part file.

Notice these important features of the FreeCAD window.

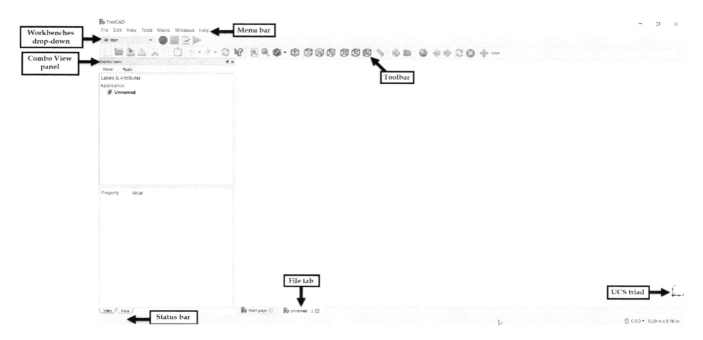

Workbenches in FreeCAD

A workbench is a set of tools and environment that can be used to create parts, assemblies, drawing, and so on. There are many workbenches available in FreeCAD. You can activate different workbenches by using the **Workbenches** drop-down located on the top-left corner.

User Interface

Various components of the user interface are discussed next.

Menu bar

Menu bar is located at the top of the window. It has various options (menu titles). When you click on a menu title, a drop-down appears. Select any option from this drop-down.

FreeCAD

File   Edit   View   Tools   Macro   Windows   Help

File Menu

The **File Menu** appears when you click on the **File** option located at the top left corner of the window. This menu contains the options to open, print, export, save, and close a file.

Toolbar

A toolbar is a set of tools, which help you to perform various operations. Various toolbars available in different workbenches are given next.

Start Toolbars	
Workbench ⮕ Start ▾	This toolbar has a drop-down to change the workbench.
Macro ● ■ ▦ ▶	This toolbar has tools to create and execute macros.
File □ 📁 ⤴ ⤵ ✂ □ ↩ ▾ ↪ ▾ 🔁 ▶?	This toolbar has tools to create, open, and save files. You can also print, cut, copy, paste, undo, redo, recompute, and seek help.
View 🔍 🔍 ◎ ▾ ⬡ ⬡ ⬡ ⬡ ⬡ ⬡ ⬡ ⬡ 📏	This toolbar has tools to manipulate the view of the model.
Structure 📦 📁	It has tools to create or open part files.

Navigation	It has tools to open a website in FreeCAD.
Sketcher Toolbars	
Sketcher	This toolbar has tools to start or exit a sketch.
Sketcher geometries	This toolbar has tools to create sketch elements.
Sketcher constraints	This toolbar has tools to apply constraints between sketch elements.
Sketcher B-spline tools	This toolbar has tools to create and edit B-splines.
Sketcher Virtual Space	This toolbar helps you to hide or show constraints.
Sketch Tools	This toolbar has various selection tools and options that aid you to create sketch elements very fast.
Part Design Toolbars	

Part Design Helper	This toolbar has tools to create new bodies, datum points, axis, datum planes, and clones. You can also create or leave a sketch.
Part Design Modeling	This toolbar has commands to create solid features based on the sketch geometry.
Assembly 2 Toolbars	
Assembly 2	This toolbar has tools to create components or insert existing components into an assembly. This toolbar also has tools to apply constraints between components.
Assembly 2 Shortcuts	This toolbar has tools to change the direction of the constraints, lock rotation, and insert multiple bolts into holes.
TechDraw Toolbars	
TechDraw Views	This toolbar has tools to generate standard views of a 3D geometry.

TechDraw Clips	It has tools to add or remove clips to the drawing sheet.
TechDraw Pages	The tools on this toolbar will help you to add a new page.
TechDraw Dimensions	The tools on this toolbar help you to add dimensions to the drawing views.
TechDraw File Access	This toolbar helps you to export the drawing page to the SVG format.
TechDraw Decoration	The tools on this toolbar help you to change the hatch patterns, insert images, and so on.

You can hide or show toolbars in the application window. To do this, click **View > Toolbars**, and then select the toolbar name from the list displayed.

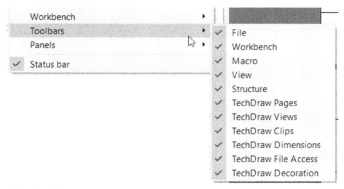

Status bar

This is available below the graphics window. It shows the action taken while using the commands.

Preselected: Pad - Unnamed.Pad.Edge4 (-4.53286, -0.020793, 10)

Model tab
Contains the list of operations carried while constructing a part.

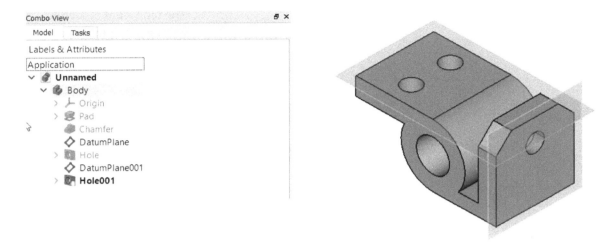

Dialog
When you click any tool in FreeCAD, the dialog related to it appears in the **Tasks** tab of the **Combo View** panel. A dialog has of various options. The following figure shows various components of a dialog.

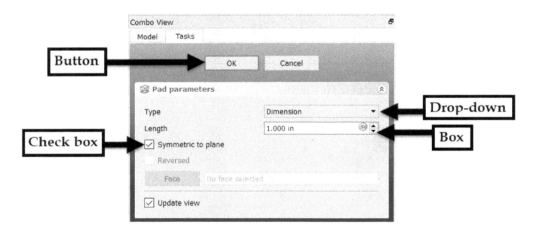

This book uses the default options on the dialog.

Navigation Styles
FreeCAD provides you with different types of mouse navigation styles: **OpenInventor**, **CAD**, **Revit**, **Blender**, **MayaGesture**, **Touchpad**, **Gesture**, and **Open Cascade**. You can select the desired navigation style from the drop-down located at the bottom right corner.

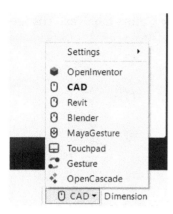

This book uses the CAD Navigation Style for your mouse. Select the **CAD** option from the **Navigation Styles** drop-down. Next, place the mouse cursor on the drop-down; the various mouse functions are displayed.

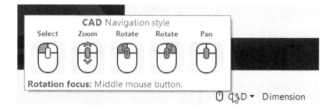

Background

To change the background color of the window, click **Edit > Preferences** on the Menu bar. On the **Preferences** dialog, click **Display** on the left side. Click the **Colors** tab and set the colors for various element types.

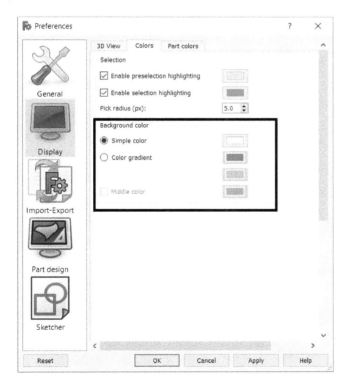

Getting Started with FreeCAD

To change the color of sketch elements, click **Sketcher** on the left side, and then click the **Colors** tab. Next, change the **Sketch colors**. Click **OK** to apply the changes.

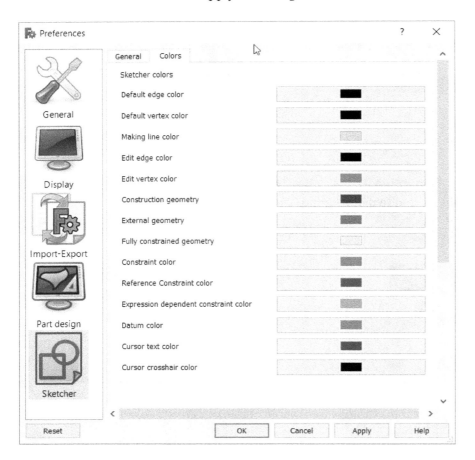

Chapter 2: Part Modeling Basics

This chapter takes you through the creation of your first FreeCAD model. You create simple parts:

In this chapter, you will:

- Create Sketches
- Create a base feature
- Add another feature to it
- Add fillets
- Shell the model
- Create Cut features

TUTORIAL 1

This tutorial takes you through the creation of your first FreeCAD model.

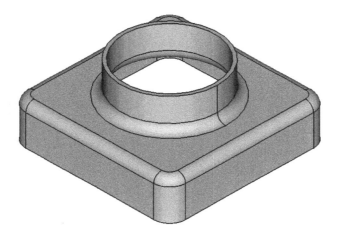

Starting a New Part File

1. To start a new part file, click **File > New** on the Menu Bar (or) click the **New** icon on the **File** toolbar.

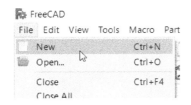

(or)

A new model window appears.

2. On the **Workbench** toolbar, select **Workbench** drop-down > **Part Design** (or) select **View > Workbench > Part Design** on the Menu bar.

Part Modeling Basics

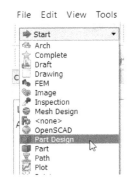

(or)

Starting a Sketch

1. To start a new sketch, click **Part Design Helper** toolbar > **Create New Sketch** (or) click **Part Design > Create Sketch** on the Menu bar.

2. Select the **XY Plane**, and then click **OK**. The sketch starts.

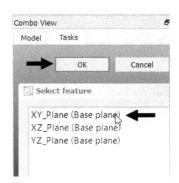

Notice that:

- The grid and sketch origin appear.
- The **Sketcher, Sketcher geometries, Sketcher constraints, Sketcher tools, Sketcher B-spline tools**, and **Sketcher virtual space** toolbars are displayed.
- "**Empty Sketch**" appears in the **Solver Messages** section in the **Combo View** panel.

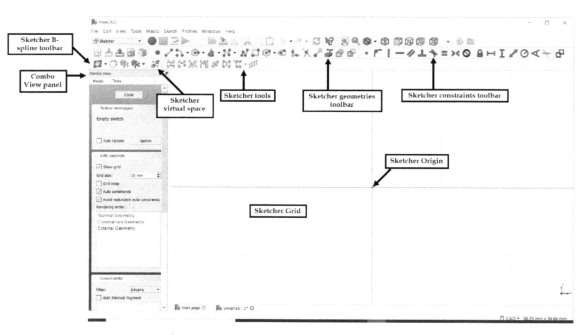

Before you begin sketching, make sure that your FreeCAD settings match the settings used in this tutorial.

3. Click **Edit > Preferences** on the Menu bar. The **Preferences** dialog box appears.
4. On the **Preferences** dialog box, click the **General** option at the left side, and then click the **Units** tab.
5. Select **User System > Standard (mm/kg/s/degree)**.
6. Type **2** in the **Number of decimal** box.
7. Click **OK**.
8. In the **Edit Controls** section of the **Task** tab of the **Combo View** panel, make sure that the **Show grid** check box is selected.

The first feature is an extruded feature from a sketched rectangular profile. You will begin by sketching the rectangle.

9. On the **Sketcher geometries** toolbar, click the **Rectangle** icon (or) click **Sketch > Sketcher geometries > Create Rectangle**.

10. Move the cursor to the sketch origin located at the center of the graphics window, and then click on it.
11. Drag the cursor towards the top right corner, and then click to create a rectangle.

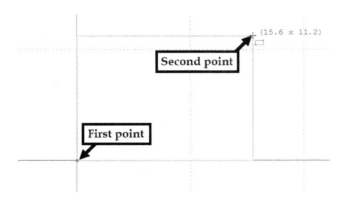

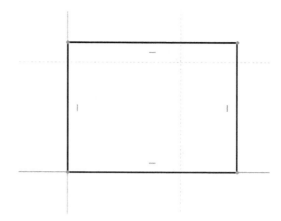

12. Press **ESC** to deactivate the tool.

Adding Constraints

In this section, you will specify the size of the sketched rectangle by adding constraints. As you add constraints, the sketch can attain any one of the following states:

Fully Constrained sketch: In a fully constrained sketch, the positions of all the entities are fully described by constraints. In a fully constrained sketch, all the entities are green color.

Under Constrained sketch: Additional constraints, are needed to completely specify the geometry. In this state, you can drag under constrained sketch entities to modify the sketch. An under constrained sketch entity is in black color.

If you add any more constraints to a fully constrained sketch, a message appears in the Solver messages section of the **Combo View** panel. It shows that constraint over constraints the sketch. In addition, it prompts you to delete anyone of the constraints. Select anyone of the constraints and press **Delete** on your keyboard. You can also convert anyone of the constraints into a reference constraint. To do this, select the constraint and click the **Toggle reference/driving constraint** icon on the **Sketcher constraints** toolbar.

1. Click **Sketcher constraints** toolbar **> Constrain vertical distance** I (or) click **Sketch > Sketcher constraints > Constrain vertical distance** on the menu bar.
2. Select the right vertical line of the rectangle.
3. Enter **100** in the **Length** box of the **Insert Length** dialog and click **OK**.
4. Click **Sketcher constraints** toolbar **> Constrain horizontal distance** (or) click **Sketch > Sketcher constraints > Constrain horizontal distance** on the menu bar.
5. Select the bottom horizontal line of the rectangle.
6. Enter **100** in the **Length** box of the **Insert Length** dialog and click the **OK** button.

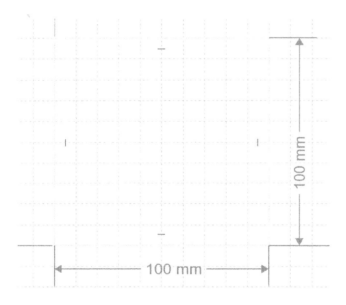

7. Press **Esc** to deactivate the **Constrain horizontal distance** tool.
8. To display the entire rectangle at full size and to center it in the graphics area, use one of the following methods:

   - Click **Fit All** on the **View** toolbar.
   - Click **View > Standard Views > Fit All** on the Menu bar.

9. Click **Close** on the **Combo View** panel.

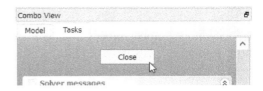

10. Again, click **Fit All** on the **View** toolbar.

Creating the Base Feature

The first feature in any part is called a base feature. You now create this feature by extruding the sketched rectangle.

1. Click **Part Design Modeling** toolbar **> Pad** (or) click **Part Design > Pad** on the Menu bar.
2. Type-in 25 in the **Length** box available on the **Pad parameters** dialog on the **Combo View** panel.
3. Click **OK** on the **Combo View** panel to create the pad feature.
4. On the **View** toolbar, click the **Axonometric** icon.

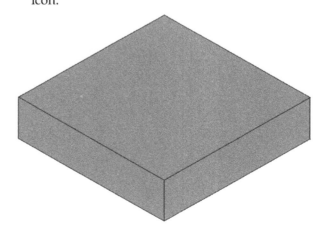

Notice the new feature, **Pad**, in the **Model** tab of the **Combo View** panel.

To magnify a model in the graphics area, you can use the zoom tools available on the **Zoom** sub menu in the **View** menu of the Menu bar.

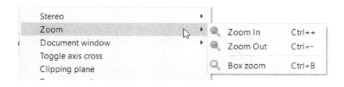

Normal Mode

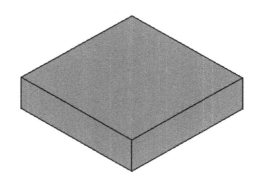

Click **Zoom In** to zoom into the model.

Click **Zoom Out** to zoom out of the model.

Click **Box Zoom**, and then drag the pointer to create a rectangle; the area in the rectangle zooms to fill the window.

Flat lines

The **View** toolbar has two more zoom tools: **Fit All** and **Fit Selection**.

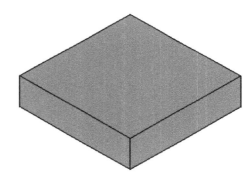

Click **Fit All** to display the part full size in the current window.

Click on a vertex, an edge, or a feature, and then click **Fit Selection** ; the selected item zooms to fill the window.

Shaded

To display the part in different draw styles, select the options in the **Draw Style** drop-down on the **View** toolbar.

Wireframe

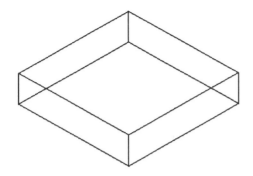

Points

Hidden Line

No Shading

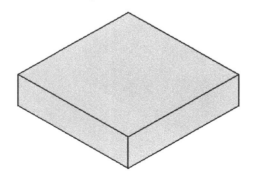

The default draw style for parts and assemblies is **Normal Mode**. You may change the draw style whenever you want.

Adding a Pad Feature

To create additional features on the part, you need to draw sketches on the model faces or planes, and then extrude them.

1. On the **Part Design Helper** toolbar, click the **Create a new datum plane** ◈ icon.
2. Click on the top face of the first feature.

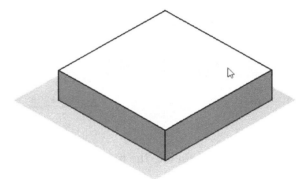

3. Click **OK** on the **Tasks** tab of the **Combo View** panel.
4. Click **Create Sketch** on the **Face tools** section of the **Tasks** tab. A new sketch is started.

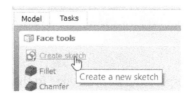

5. Click on the datum plane.
6. On the **Sketcher geometries** toolbar, click **Circle** drop-down > **Center and rim point**.

7. Click to specify the center point of the circle.
8. Move the pointer outward and click to create the circle.

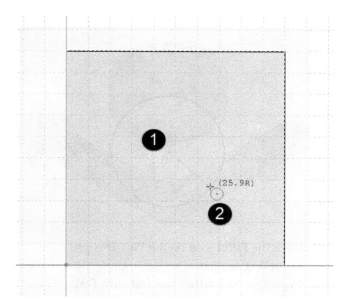

9. On the **Sketcher constraints** toolbar, click the **Constrain distance** icon.
10. Select the center point of the circle.
11. Select the horizontal axis of the sketch.
12. Type **50** in the **Length** box of the **Insert Length** dialog.
13. Click **OK**.

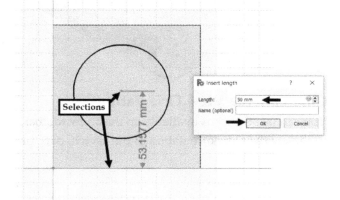

14. Select the center point of the circle and the vertical axis of the sketch.
15. Type **50** in the **Length** box and click **OK**.

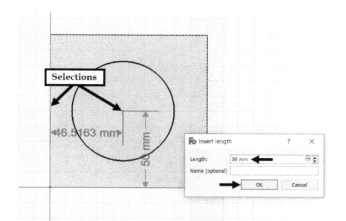

16. On the **Sketcher constraints** toolbar, click the **Constrain radius** icon.
17. Select the circle and type **30** in the **Radius** box of the **Change Radius** dialog.

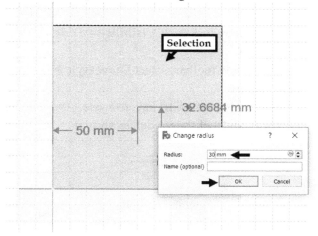

18. Click **OK**.
19. On the **Sketcher** toolbar, click the **Leave Sketch** icon.
20. On the **Part Design Modeling** toolbar, click the **Pad** icon.
21. Type **20** in the **Length** box of the **Pad Parameters** dialog.
22. Click **OK**.

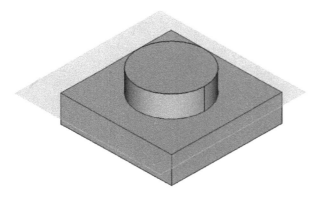

Filleting the Corners

The **Fillet** tool allows you to fillet the corners of the model.

1. Click on the datum plane displayed on the model.
2. On the Menu bar, click **Visibility > Hide Selection**.
3. On the **View** toolbar, select **Draw style > Wireframe**.
4. Press and hold the Ctrl key and select the vertical edges of the model, as shown.

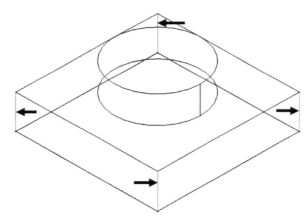

5. On the **Part Design Modeling** toolbar, click the Fillet icon.
6. Type **10** in the **Radius** box.
7. Click **OK**.
8. On the **View** toolbar, select **Draw style > Flat lines**.
9. Click on the horizontal face of the model, as shown.

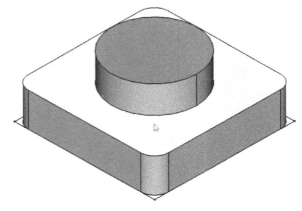

10. Click the **Fillet** icon on the **Part Design Modeling** toolbar.
11. Type **5** in the **Radius** box and click **OK**.

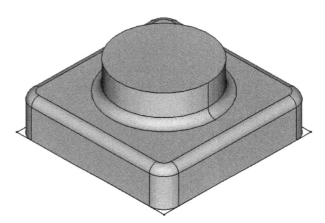

Changing the View Orientation

You can use the icons available on the **View** toolbar to change the view orientation of the sketch, part, or assembly.

- Front
- Top
- Right
- Rear
- Bottom
- Left
- Axonometric

The default planes of the part corresponding to the standard views are as follows:

- **XZ Plane - Front** or **Back**
- **YZ Plane - Top** or **Bottom**
- **XY Plane - Right** or **Left**

Rotating and Moving the Part

In addition to standard views, you can view the model from different angles by rotating them. By doing so you can select the hidden faces and edges easily.

To rotate the part, use one of the following methods:

- On the Menu bar, click **View > Standard Views > Rotate Left** to rotate the model towards left.
- On the Menu bar, click **Views > Standard Views > Rotate Right** to rotate the model towards right.
- Press and hold middle and the right mouse button. Next, drag the cursor to rotate the model.
- To rotate the part in 90° increments, press and hold the **Shift** key and use the arrow keys.

To move the part view, use one of the following methods:

- Press and hold the middle mouse button and drag the cursor.
- Press and hold the **Alt** or **Ctrl** key and use the arrow keys to move the view up, down, left, or right.

Shelling the model

The **Thickness** tool allows you to shell the model.

1. Press and hold the middle and right mouse buttons.
2. Move the cursor upward to display the bottom face.
3. Click on the bottom face.

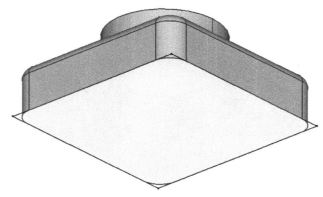

4. On the **Part Design Modeling** toolbar, click the **Thickness** icon.
5. In the **Thickness Parameters** dialog, select **Mode > Skin**.
6. Select **Join Type > Intersection**.
7. Type **2** in the **Thickness** box.
8. Check the **Make thickness inwards** option.
9. On the **View** toolbar, click the **Axonometric** icon.
10. On the **Thickness parameters** dialog, click the **Add face** button.
11. Select the top face of the second feature.

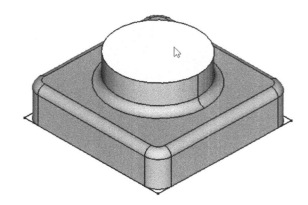

12. Click **OK**.

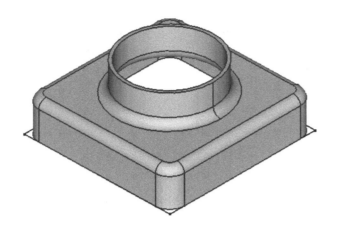

Saving the Part

1. Click **File > Save** on the Menu bar.
2. On the **Save As** dialog, type-in **Tutorial1** in the **File name** box.
3. Click **Save** to save the file.
4. Click **File Menu > Close**.

Note:

*.FCStd is the file extension for all the files that you create in FreeCAD.

TUTORIAL 2

In this tutorial, you create the part shown below.

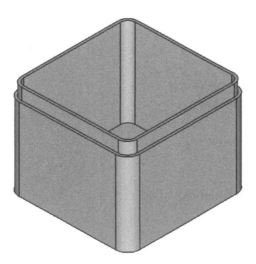

Starting a New File

1. On the Menu bar, click **File > New**.
2. Select **Part Design** from the **Workbench** drop-down.

Sketching for the Pad Feature

1. Click **Create Sketch** icon on the **Part Design Helper** toolbar.
2. Select the XY plane.
3. Click **OK**.

4. Click **Rectangle** 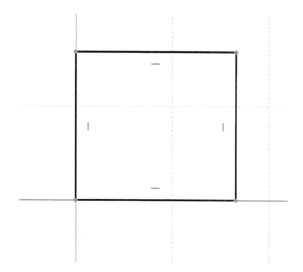 on the **Sketcher geometries** toolbar.
5. Create a rectangle, shown in figure.

6. Add vertical and horizontal distance constraints to the rectangle, as shown.

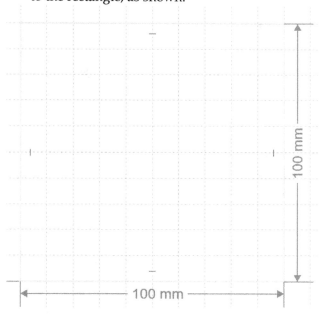

7. Click the **Leave Sketch** icon on the **Sketcher** toolbar.
8. Click **Pad** on the **Create** panel.
9. Type **80** in the **Length** box and click **OK**.
10. Click the **Axonometric** icon on the **View** toolbar.

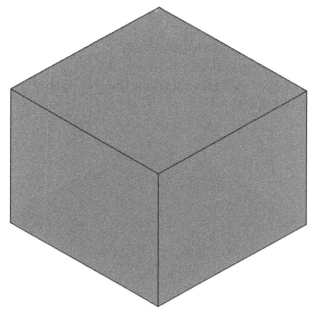

11. Select **Draw Style > Wireframe** on the **View** toolbar.
12. Press and hold the Ctrl key and select the vertical edges.

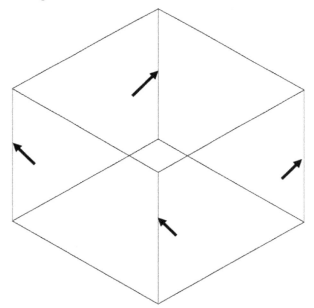

13. On the **Part Design Modeling** toolbar, click the **Fillet** icon.
14. Type 10 in the **Radius** box.
15. Click **OK**.
16. On the **View** toolbar, select **Draw Style > Flat lines**.
17. Click on the top face of the model.

18. Click the **Thickness** ⬛ icon on the **Part Design Modeling** toolbar.
19. Type 4 in the **Thickness** box.
20. Select **Join Type > Intersection**.
21. Check the **Make thickness inwards** option.
22. Click **OK**.

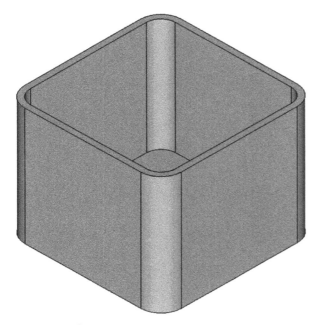

Creating the Cut Feature

1. On the **Part Design Helper** toolbar, click the **Create a new datum plane** ◇ icon.
2. Click on the top face of the first feature.

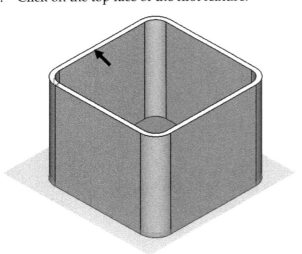

3. Click **OK** on the **Tasks** tab of the **Combo View** panel.

4. Click **Create Sketch** on the **Face tools** section of the **Tasks** tab. A new sketch is started.

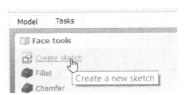

5. Click on the datum plane.
6. Click the **External geometry** 🖌 icon on the **Sketcher geometries** toolbar.
7. Click on the inner edges of the model.

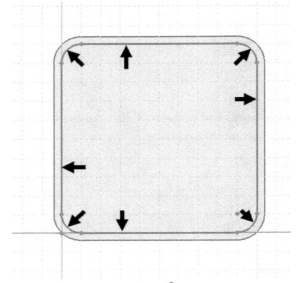

8. Click the **Rectangle** ▢ icon on the **Sketcher geometries** toolbar.
9. Create a rectangle, as shown.

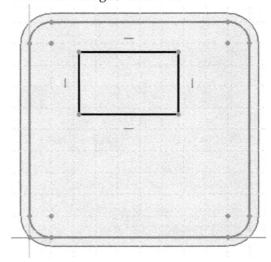

10. Click the **Create fillet** icon on the **Sketcher geometries** toolbar.
11. Click on the left vertical and the top horizontal line.

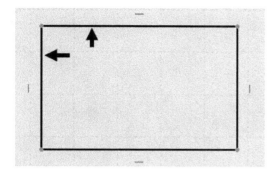

A fillet is created at the corner.

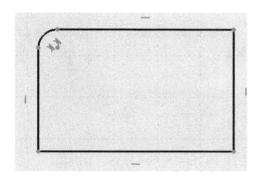

12. Likewise, create fillets at the remaining corners, as shown.

13. Click the **Constrain equal** icon on the **Sketcher constraints** toolbar.
14. Select the top -right and top-left fillets of the rectangle; the fillets are made equal.

15. Select the top-left and bottom-left fillets of the rectangle.
16. Select the bottom-left and the bottom-right fillets of the rectangle.

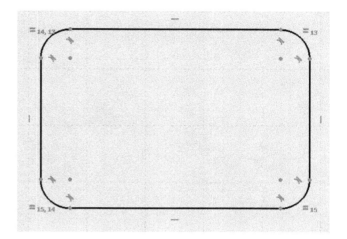

17. Click the **Constrain Coincident** icon on thee **Sketcher constraints** toolbar.
18. Select the center point of the top right fillet of the rectangle.
19. Select the top right fillet of the external geometry.

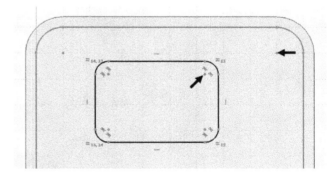

20. Click the bottom-left fillet of the rectangle.
21. Click the bottom-left fillet of the external geometry.

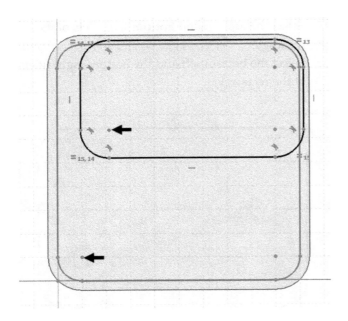

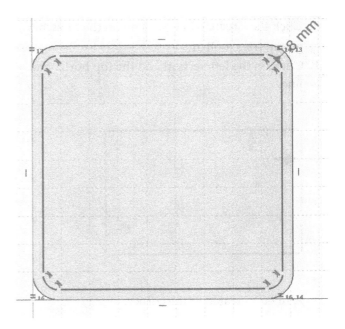

The centerpoints of the fillets are coincident to each other.

26. Likewise, create another rectangle with fillets, as shown.

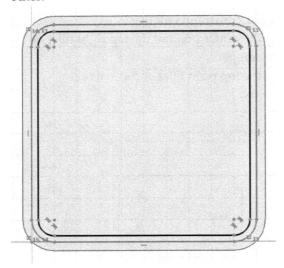

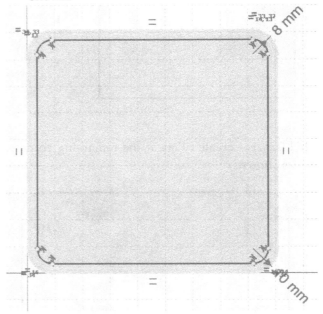

22. Click **Constrain radius** ⊘ on the **Sketcher constraints** toolbar.
23. Select anyone of the fillets of the rectangle.
24. Type 12 in the **Radius** box.
25. Click **OK**.

27. Click **Leave Sketch** on the **Sketcher** toolbar.
28. Click the **Pocket** 🪟 icon on the **Part Design Modeling** toolbar.
29. Type **10** in the **Length** box in the **Pocket parameters** dialog.
30. Click **OK**.

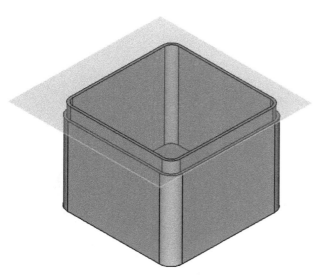

31. Save and close the file.

Chapter 3: Assembly Basics

In this chapter, you will:

- Add Components to assembly
- Apply constraints between components

TUTORIAL 1

This tutorial takes you through the creation of your first assembly.

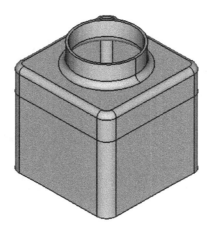

Starting a New Assembly File

1. Open the FreeCAD application.
2. Click **Tools > Addon Manager**.
3. On the **Addon Manager** dialog, click the Workbenches tab.
4. Select **assembly2** from the list.
5. Click **Install/Update**.
6. Click **Close**.
7. Close the FreeCAD application, and then restart it.
8. Click **File > New** on the Menu bar.
9. Select **Workbenches** drop-down > **Assembly 2**.

Inserting the Base Component

1. To insert the base component, click **Import a part from another FreeCAD document** icon on the **Assembly 2** toolbar.

2. Browse to the location of the **Tutorial 2** file of Chapter 1, and then double-click on it.
3. Click the **Axonometric** icon on the **View** toolbar.

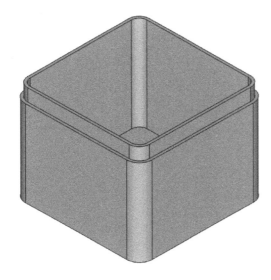

Adding the second component

1. To insert the second component, click **Assembly2 > Import a part from another FreedCAD document** on the Menu bar.
2. Browse to the location of the **Tutorial** 1 file of Chapter 1, and then double click on it.
3. Click in the window to place the component.

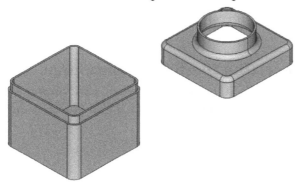

Applying Constraints

After adding the components to the assembly environment, you need to apply constraints between them. By applying constraints, you establish relationships between components.

The **Assembly 2** toolbar has various tools to apply constraints between the components.

Different assembly constraints that can be applied are given next.

Circular edge Constraint: This constraint is used to make two circular edges concentric.

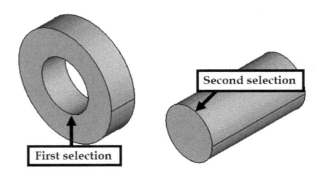

Click the **Flip** icon on the **Assembly 2 Shortcuts** toolbar to reverse the direction.

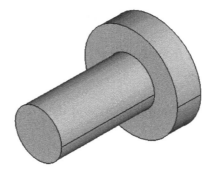

Click the **Lock Rotation** icon on the **Assembly 2 Shortcuts** toolbar, if you want to lock the rotation of the component.

Plane Constraint: Using this constraint, you can make two planar faces coplanar to each other.

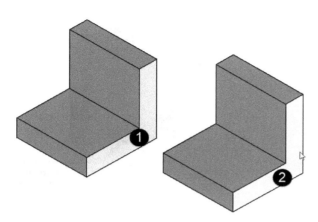

You can change the direction in which the planar faces touch each other. To do this, click the **Flip direction** icon on the **Assembly 2 shortcuts** toolbar.

You can also change the direction by using the **Combo View** panel. To do this, select the **planeConstraint** from the **Model** tab of the **Combo View** panel.

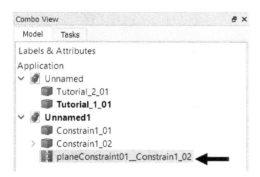

Next, select an option from the **direction** drop-down located in the **Constraint info** section.

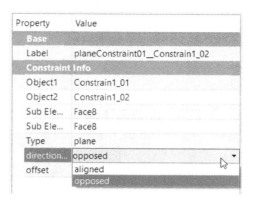

Property	Value
Base	
Label	planeConstraint01_Constrain1_02
Constraint Info	
Object1	Constrain1_01
Object2	Constrain1_02
Sub Ele...	Face8
Sub Ele...	Face8
Type	plane
direction...	opposed
offset	aligned
	opposed

Aligned

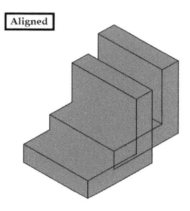

After changing the direction option, click the **Update parts imported into the assembly** icon on the **Assembly 2** toolbar.

Opposed

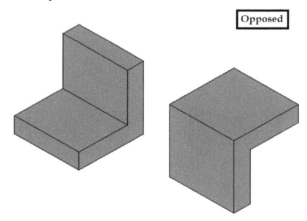

Axial Constraint: This constraint allows you to align the centerlines of the cylindrical faces. Select the two cylindrical faces to be aligned.

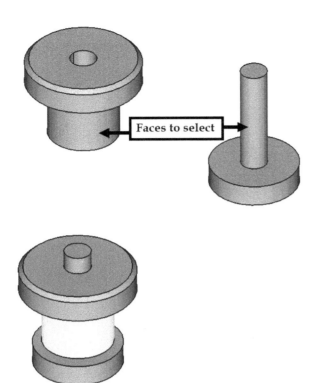

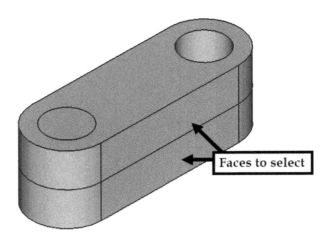

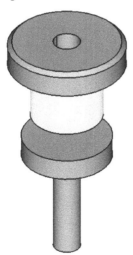

Click the **Flip** icon on the **Assembly 2 shortcuts** toolbar; the axes of the selected cylindrical faces will be positioned in the direction opposite to each other.

Select the **angleConstraint** from the **Model** tab in the **Combo View** panel. In the **Constraints info** section, type 45 in the **angle** box.

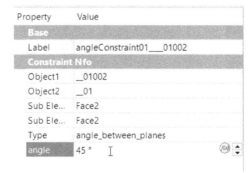

Angle Constraint: Applies the angle constraint between two components.

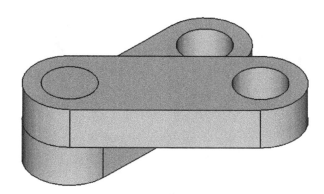

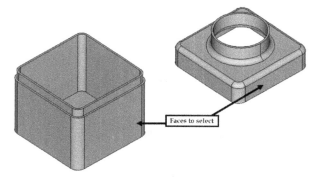

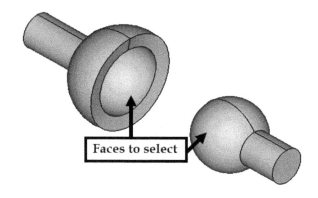

3. On the Menu bar, click **Assembly 2 > Add a plane constraint**.
4. Select the planar faces of the two parts, as shown.

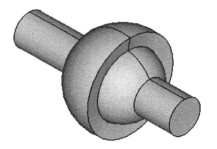

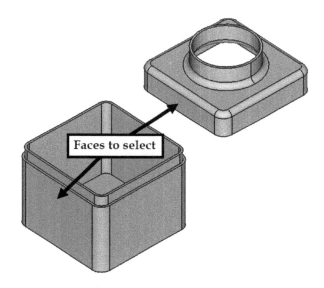

Spherical Surface Constraint: This constraint is used to align two spherical surfaces.

5. Click the **Animate degrees of freedom** icon on the **Assembly 2** toolbar. The degrees of freedom is animated.

1. Click the **Add a plane constraint between two objects** icon on the **Assembly 2** toolbar.
2. Click on the planar faces of the two parts, as shown.

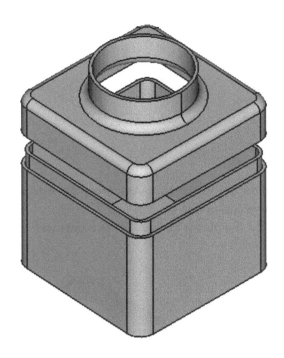

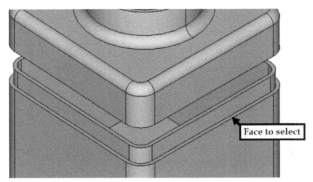

8. Select the second face, as shown.

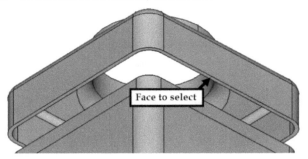

The assembly is constrained fully.

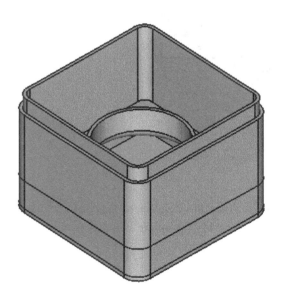

You need to remove this degree of freedom.

6. Click the **Add a plane constraint between two objects** icon on the **Assembly 2** toolbar.

7. Select the first face.

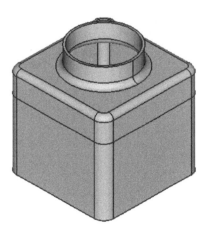

9. Click the **Check assembly for part overlap** icon on the **Assembly 2** toolbar. The Passed message box appears.

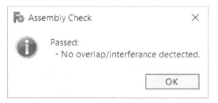

10. Click **File > Save** on the Menu bar.

Assembly Basics

11. Type **Assembly_tutorial** in the **File name**.
12. Click **Save**.
13. Click **File > Close** on the Menu bar.

Chapter 4: Sketching

In this chapter, you will learn the sketching tools. You will learn to create:

- Polylines
- Polygons
- Slots
- Constraints
- B-Splines
- Ellipses
- Circles
- Trim
- Extend
- Toggle Construction geometry

Creating Polylines

The **Create Polyline** tool is the most commonly used tool while creating a sketch.

1. To activate this tool, you need to click the **Create Polyline** 🖋 icon on the **Sketcher geometries** toolbar.
2. Click in the graphics window and move the pointer.
3. Click to create a line. Notice that another line is attached to the cursor.

4. Move the pointer and click to create another line.

5. Press the **M** key on your keyboard.
6. Move the pointer and click to create a line perpendicular to the previous line.

7. Press the **M** key twice on your keyboard.
8. Move the pointer and click to create a line colinear to the previous line.

9. Press the **M** key thrice on your keyboard; a grey arc tangent to the previous line appears.

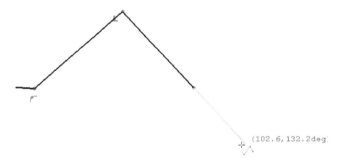

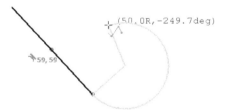

10. Move the pointer and click to create an arc tangent to the previous line.

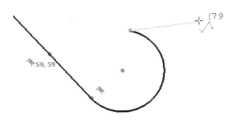

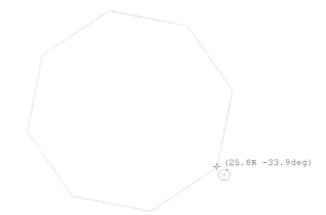

11. Press the M key four times to display an arc normal to the previous line.
12. Again, press the M key to change the direction of the normal arc.

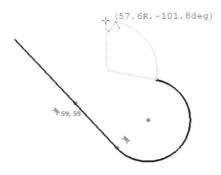

13. Click to create the normal arc.
14. Right click to end the line chain.
15. Again, right click to deactivate the **Create Polyline** tool.

Creating Polygons

A Polygon is a shape having many sides ranging from 3 to 1024. In FreeCAD, you can create regular polygons having sides with equal length. Follow the steps given next to create a polygon.

1. Click the **Create Sketch** icon on the **Sketcher** toolbar.
2. On the **Sketcher geometries** toolbar, select **Polygon > Regular Polygon**.
3. On the **Create array** dialog, type **8** in the **Number of Sides** box.
4. Click **OK**.
5. Click to define the center of the polygon.
6. Move the pointer and click to define the size and angle of the polygon.

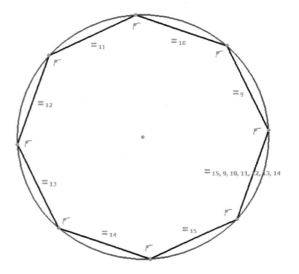

Creating Slot

The **Create Slot** tool is used to create slots.

1. Click the **Create Slot** icon on the **Sketcher geometries** toolbar.
2. Click to specify the centerpoint of the first semicircle.
3. Move the pointer and click to specify the endpoint of the second semicircle.

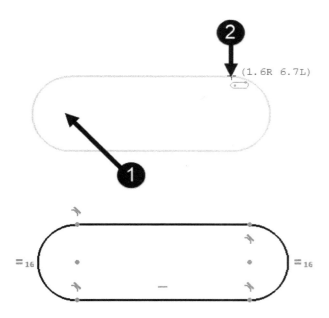

Constraints

Constraints are used to control the shape of a sketch by establishing relationships between the sketch elements. You can add constraints using the tools available on the **Sketcher constraints** toolbar.

Constrain Coincident

This constrain connects a point to another point.

1. On the **Sketcher constraints** toolbar, click **Constrain Coincident** •.
2. Select two points. The selected points are connected together.

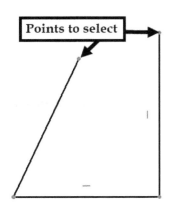

— **Constrain horizontally**

To apply the **Horizontal** constraint, click on a line and click the **Constrain Horizontally** icon on the **Sketcher constraints** toolbar.

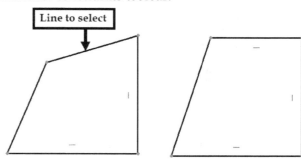

| Constrain Vertically

Use the **Constrain Vertically** icon to make a line vertical.

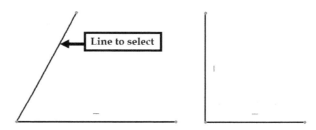

⚓ Constrain tangent

This constraint makes an arc, circle, or line tangent to another arc or circle. Click the **Constrain Tangent** icon on the **Sketcher constraints**. Select a circle, arc, or line. Next, select another circle, or arc; the two elements will be tangent to each other.

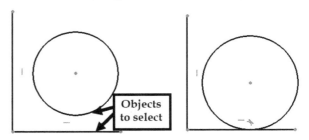

⫽ Constrain Parallel

Use the **Constrain Parallel** icon to make two lines parallel to each other. To do this, click the **Constrain Parallel** icon on the **Sketcher constraints**. Next, select the two lines to be parallel.

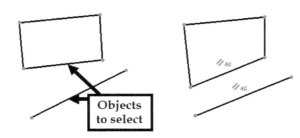

Constrain Perpendicular

Use the **Constrain Perpendicular** icon to make two entities perpendicular to each other.

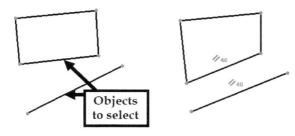

Auto Constraints

FreeCAD automatically adds constraints when you create sketch elements.

1. Start a new sketch and activate the **Create Polyline** tool from the **Sketcher geometries** toolbar.
2. Click to specify the start point of the line.
3. Move the pointer in the horizontal direction and notice the **Horizontal** constraint flag.
4. Click to create a line with the **Horizontal** constraint.

5. Move the pointer vertically in the upward direction and notice the **Vertical** constraint flag.
6. Click to create a line with the **Vertical** constraint.

7. Create an inclined line as shown.

8. Press **Esc** twice to deactivate the tool.
9. On the **Sketcher geometries** toolbar, click the **Create Circle** icon, and then create a circle.
10. Activate the **Create Line** tool and click on the circle.
11. Move the pointer around the circle and notice that the line maintains the **Tangent** constraint with the circle.

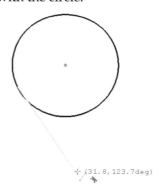

12. Click to create a line which is tangent and coincident to the circle. Right click to deactivate the **Create Line** tool.

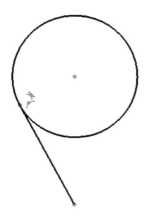

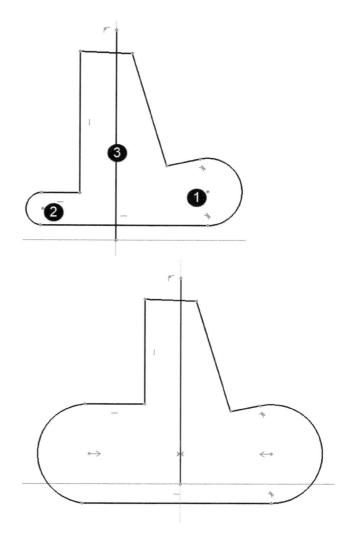

Deleting Constraints

You can delete constraints by using the following methods.

- Select the constraint and press **Delete** on your keyboard.

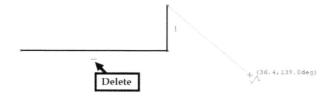

Constrain symmetrical

Use the **Constrain symmetrical** tool to make two sketch elements symmetric about a centerline.

1. Click on the elements to make symmetric.
2. Click on the symmetric line.
3. Click the **Constrain symmetrical** ✕ icon on the **Sketcher constraints** toolbar.

Constrain Lock

This constrain locks a selected point by adding dimensions to it.

1. On the **Sketcher constraint** toolbar, click the **Constrain Lock** 🔒 icon.
2. Select a point from the sketch.

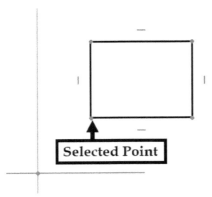

Dimensions are created between the selected point and the origin.

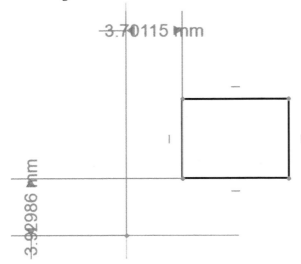

Constrain Block

This constraint fixes the sketch object at its location.

1. Click the **Constrain Block** 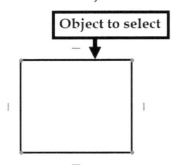 icon on the **Sketcher constraints** toolbar.
2. Select an object from the sketch.

3. Click the object an notice that it is fixed at its location.

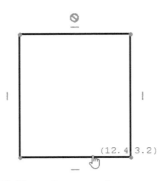

Hiding Constraints

To hide sketch constraints, uncheck the check box next to it in the **Constraints** section located in the **Combo View** panel.

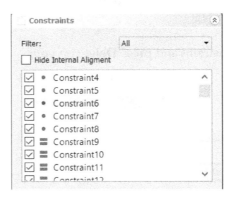

Create B-Spline

This command creates a smooth B-spline curve using the control points you select.

1. On the **Sketcher geometries** toolbar, click **B-Spline** drop-down > **Create B-Spline**.
2. Click to define points in the graphics window.

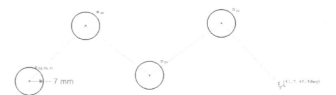

3. Right click to create a spline controlled by the selected points.
4. Press Esc to deactivate this command.

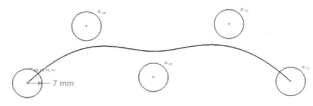

Create periodic B-spline

If you want to create a closed B-spline, then click the **B-spline** drop-down > **Create periodic B-spline** on the **Sketcher geometries** toolbar. Next, specify the control points. Right click to create a closed B-spline.

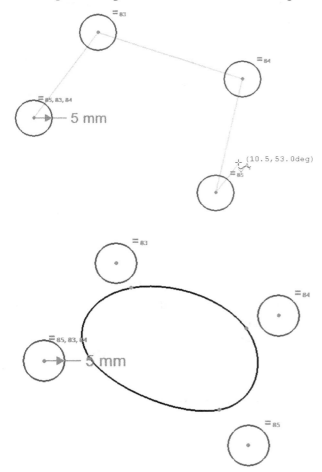

Ellipses

Ellipses are also non-uniform curves, but they have a regular shape. They are actually splines created in regular closed shapes.

1. Click the **Create Sketch** icon on the Sketcher toolbar.

2. Select anyone of the datum planes from the **Combo View** panel.

3. Click **OK**.

4. On the **Sketcher geometries** toolbar, click **Conic** drop-down > **Create ellipse by center** .

5. Pick a point in the graphics window to define the center of the ellipse.

6. Move the pointer and click to define the radius and orientation of the first axis.

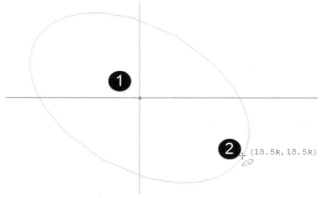

7. Move the pointer and click to define the radius of the second axis.

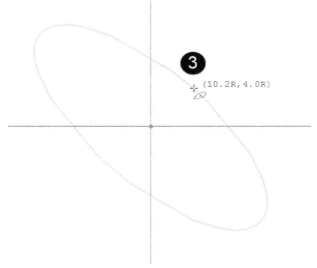

Click and drag the ellipse to notice that it is under-defined. You need to add dimensions and constrains to fully define the ellipse.

8. On the **Sketcher constraints** toolbar, click **Constrain angle** .

9. Select the lines, as shown.

41

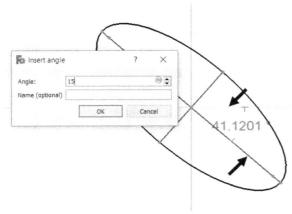

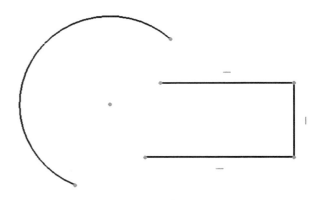

10. Type 15 in the **Angle** box and click **OK**.

11. Click the **Constrain distance** icon on the **Sketcher constraints** toolbar.

12. Select the major axis line.

13. Type 25 in the **Length** box and click the **OK** button.

2. Click the **Extend edge** icon on the **Sketcher geometries** toolbar.

3. Select the horizontal open line.

4. Select the arc. This will extend the line up to the arc.

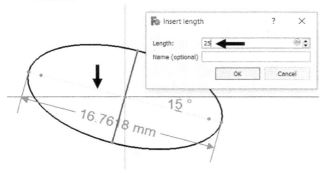

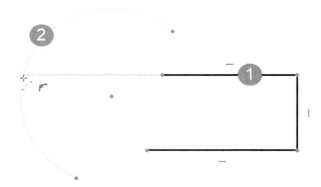

14. Likewise, add the distance constrain to the minor axis. This fully-defines the sketch.

Likewise, extend the other elements, as shown.

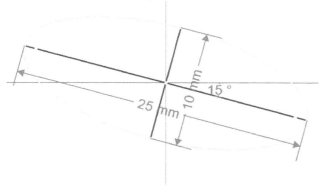

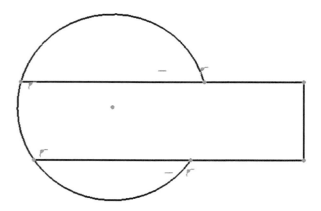

Extend Edge

The **Extend Edge** tool is used to extend lines, arcs and other open entities to connect to other objects.

1. Create a sketch as shown below.

Trim Edge

The **Trim Edge** tool is used to trim the unwanted portions of the sketch using an intersecting edge.

1. Click the **Trim Edge** icon on the **Sketcher geometries** toolbar.
2. Click on the edges to trim.

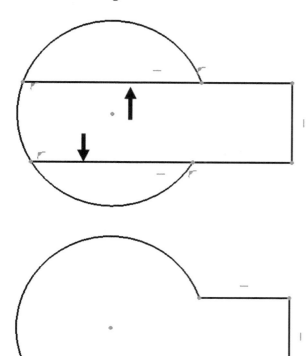

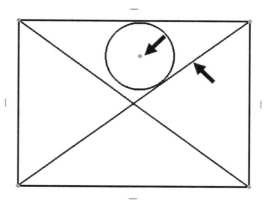

The point is made coincident with the object. Likewise, make the point coincident with another object, as shown.

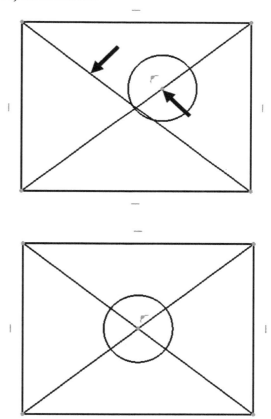

Constrain point onto object

This constrain makes a point coincident with a sketch element.

1. Click the **Constrain point onto object** icon on the **Sketcher constraints** toolbar.
2. Select the point and object, as shown.

Toggle Construction geometry

This command converts a standard sketch element into a construction element. Construction elements support you to create a sketch of a desired shape and size. To convert a standard sketch element to construction element, click on it and select **Toggle Construction geometry** on the **Sketcher geometries** toolbar.

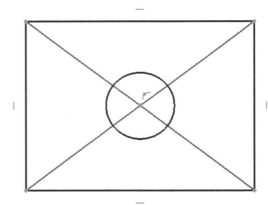

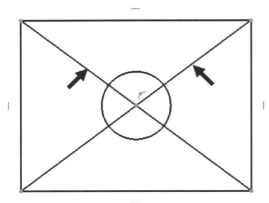

You can also convert it back to a standard sketch element by clicking on it and selecting **Toggle Construction geometry** icon on the **Sketcher geometries** toolbar.

Chapter 5: Additional Modeling Tools

In this chapter, you create models using additional modeling tools. You will learn to:

- Create slots
- Create circular patterns
- Create holes
- Create chamfers
- Create shells
- Create coils
- Create a loft feature
- Create a sweep feature

TUTORIAL 1

In this tutorial, you create the model shown in figure:

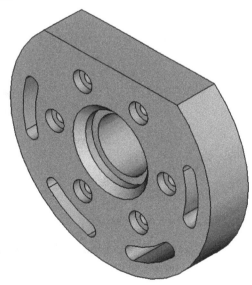

Creating the First Feature

1. Open the FreeCAD application.
2. Click **File > New** on the Menu bar.
3. Select the **Part Design** option from the **Workbenches** drop-down.
4. On the Menu bar, click **Edit > Preferences**.
5. Click the **Units** tab on the **Preferences** dialog.
6. Select **User system > Imperial decimal**.

7. Type **3** in the **Number of decimals** box.
8. Click **OK**.
9. Click the **Create sketch** icon on the **Part Design Helper** toolbar, and then select the XZ Plane.
10. Click **OK** to start the sketch.
11. Click the **Create Circle** icon on the **Sketcher geometries** toolbar.
12. Select the origin point of the sketch.
13. Move the pointer outward and click to create a circle.

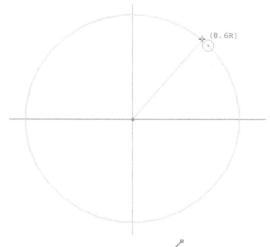

14. Click the **Create Line** icon on the **Sketcher geometries** toolbar.
15. Specify a point at the location outside the circle, as shown.

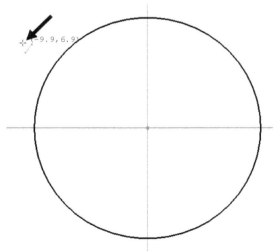

16. Move the pointer horizontally and notice the Horizontal constraint symbol.

17. Click outside the circle. Press Esc to deactivate the **Create Line** tool.

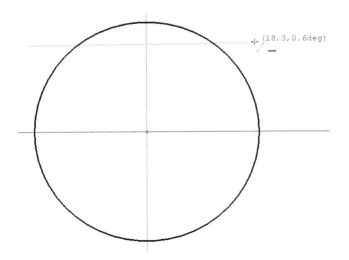

18. Click the **Trim Edge** icon on the **Sketcher geometries** toolbar.
19. Click on the portions of the sketch to be trimmed, as shown below.

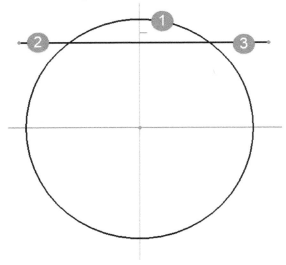

20. Add the **Radius** and **Horizontal Distance** constraints to the sketch.

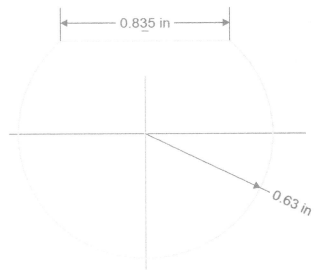

21. Click the **Close** button on the **Combo View** panel.
22. Click the **Pad** icon on the **Part Design Modeling** toolbar.
23. Type **0.236** in the **Length** box and click **OK**.

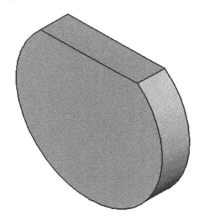

Creating the Pocket feature

1. Click the **Create a datum plane** icon on the **Part Design Helper** toolbar.
2. Click on the front face of the model, and then click **OK**.

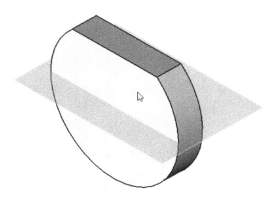

3. Click the **Create Sketch** icon on the **Part Design Helper** toolbar.
4. On the **Sketcher geometries** toolbar, click **Arc drop-down > Center and end points**.
5. Select the origin as the center point.
6. Move the cursor outside and click in the first quadrant of the circle to specify the start point of the arc.
7. Move the cursor and click in the fourth quadrant of the circle to specify the end point of the arc.

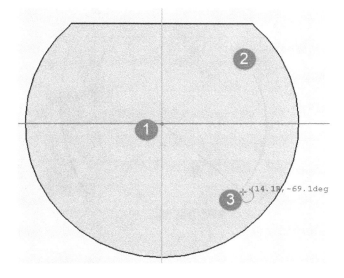

8. Likewise, create another centerpoint arc, as shown.

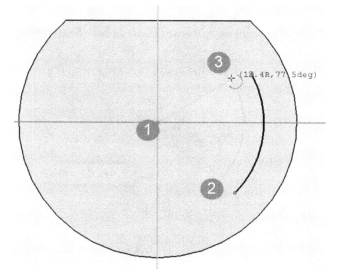

9. On the **Sketcher geometries** toolbar, click **Arc drop-down > End points and rim point**.
10. Zoom to the first quadrant.
11. Select the end point of the first and second arcs, as shown.
12. Move the pointer outward and click to specify the rim point of the arc.

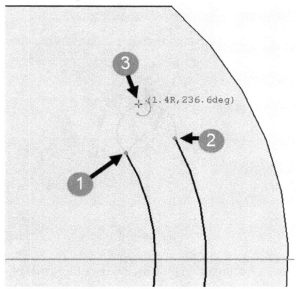

13. Use the **Constrain Coincident** tool and connect the end points of the small arc to the end points of the centerpoint arcs, if they are not properly connected.

14. Click the **Constrain Tangent** icon on the **Sketcher constraints** toolbar.

15. Select the small arc and anyone of the centerpoint arcs; the small arc is made tangent to the centerpoint arc.
16. Likewise, make the small arc tangent to the other center point arc.

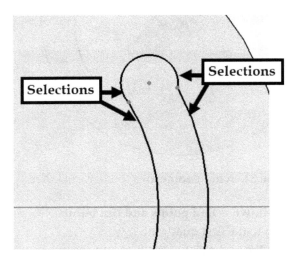

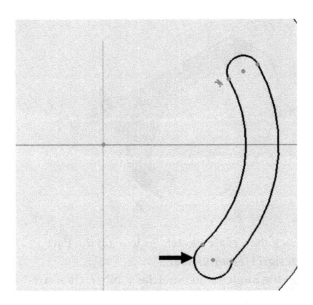

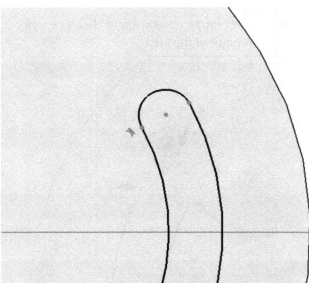

17. Likewise, create a small arc on the other ends of the centerpoint arcs.

18. Use the **Constrain Coincident** tool and connect the end points of the small arc to the end points of the centerpoint arcs, if they are not properly connected.
19. Make the small arc tangent to the centerpoint arcs.

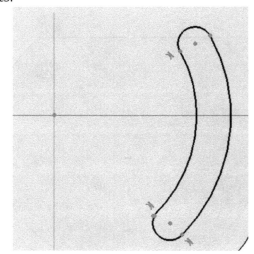

20. Click the **Toggle Construction geometry** icon on the **Sketcher geometries** toolbar.
21. Click the **Create Line** icon on the **Sketcher geometries** toolbar.
22. Select the origin point of the sketch.
23. Select the centerpoint of the small arc.

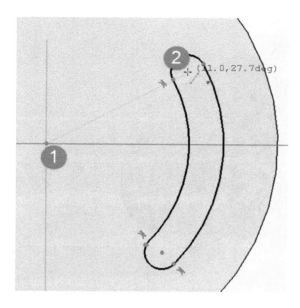

24. Likewise, create a line by selecting the origin and center point of another small arc.

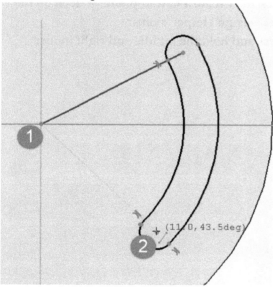

25. Use the **Constrain Coincident** tool and connect the end point of line and the center point of the small arc, if not properly connected.

26. Click the **Toggle construction geometry** icon on the **Sketcher geometries** toolbar.

27. Click the **Constrain distance** icon on the **Sketcher constraints** toolbar.

28. Select anyone of the lines.

29. Type 0.512 in the **Length** box and click **OK**.

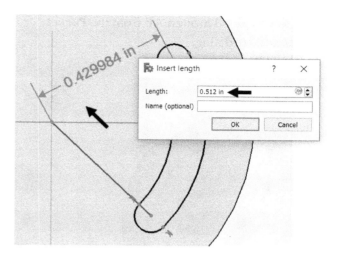

30. Click the **Constrain Radius** icon on the **Sketcher constraints** toolbar.

31. Select anyone of the small arcs.

32. Type 0.039 in the **Radius** box, and then click **OK**.

33. Click the **Constrain angle** icon on the **Sketcher constraints** toolbar.

34. Select the two lines.

35. Type **30** in the **Angle** box, and then click **OK**.

36. Select the horizontal axis of the sketch and anyone of the lines.

37. Type **15** in the **Angle** box, and then click **OK**.

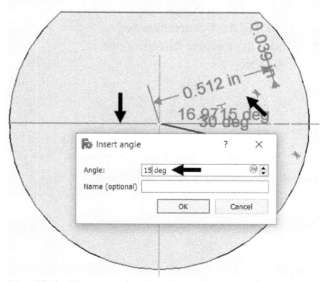

38. Click **Close** on the **Combo View** panel.

39. Click the **Pocket** icon on the **Part Design Modeling** toolbar.

40. Select **Type > Through All** from the **Pocket Parameters** dialog.
41. Click **OK**.

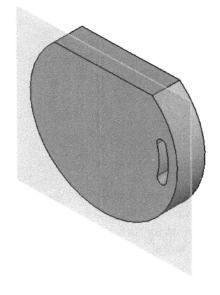

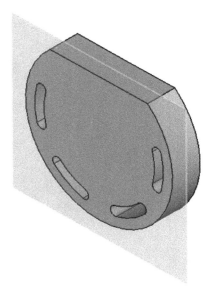

Creating a Polar Pattern

1. Click the **Polar Pattern** icon on the **Part Design Modeling** toolbar.
2. Select the **Pocket** feature from the **Select Feature** section in the **Combo View** panel.
3. Click **OK**.
4. Select **Axis > Normal sketch axis**.
5. Type **180** in the **Angle** box.
6. Type **4** in the **Occurrences** box.
7. Check the **Reverse direction** option.
8. Click **OK**.

Adding the Pad feature

1. Click the **Create a datum plane** icon on **the Part Design Helper** toolbar.
2. Press and hold the middle and right mouse

button.

3. Drag the pointer to rotate the model.

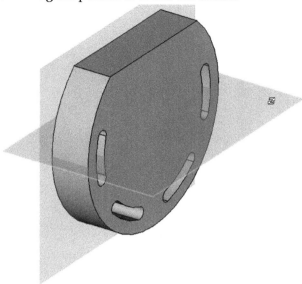

4. Click on the back face of the model.
5. Click **OK**.

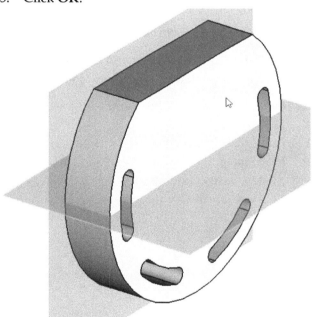

6. Click the **Create Sketch** icon on the **Part Design Helper** toolbar.
7. Select the newly created datum plane.
8. Click **OK**.
9. Click the **Create Circle** icon on the **Sketcher geometries** toolbar.
10. Select the sketch origin.
11. Move the pointer outward and click to create the

circle.

12. Click the **Constrain Radius** icon on the **Sketcher constraints** toolbar.
13. Type **0.236** in the **Radius** box and click **OK**.

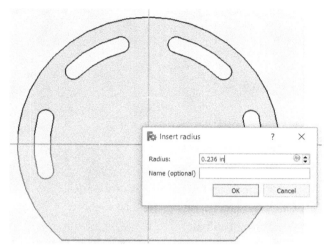

14. Click **Close** on the **Combo View** panel.
15. Click the **Pad** icon on the **Part Design Modeling** toolbar.
16. Type **0.078** in the **Length** box.
17. Click **OK**.

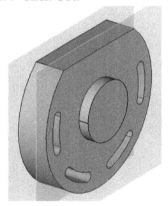

Creating a Counterbore Hole

In this section, you will create a counterbore hole concentric to the cylindrical face.

1. Click the **Create Sketch** icon on the **Part Design Helper** toolbar.
2. Select the datum plane created on the front face of the model.
3. Click **OK** on the **Combo View** panel.
4. Click the **Create Circle** icon on the **Sketcher**

geometries toolbar.

5. Select the origin point of the sketch.
6. Move the pointer outward and click to create a circle.

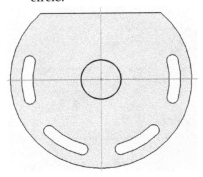

7. Click **Close** on the **Combo View** panel.
8. Click the **Hole** icon on the **Part Design Modeling** toolbar.
9. Type **8** in the **Diameter** box (You can enter the hole parameter in millimetres only).
10. Select **Depth > Through All**.
11. On the **Hole Parameters** section, under **Hole cut**, select **Type > Counterbore**.
12. Type **10** in the **Diameter** box.
13. Type **2** in the **Depth** box.
14. Under **Drill point**, select **Type > Flat**.

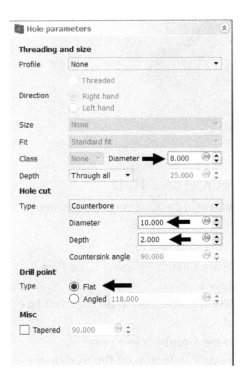

15. Click **OK** on the **Combo View** panel; the counterbore hole is created.

Creating Threaded holes

In this section, you will create a threaded hole.

1. Click the **Create Sketch** icon and select the datum planne displayed on the front face of the model.
2. Click **OK** on the **Combo View** panel.
3. Click the **Toggle Construction geometry** icon on the **Sketcher geometries** toolbar.
4. Click the **Create Line** icon on the **Sketcher geometries** toolbar.
5. Select the sketch origin.
6. Move the pointer toward right and click to create a horizontal line.
7. Likewise, create five more lines, as shown.

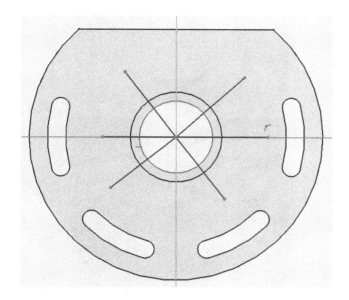

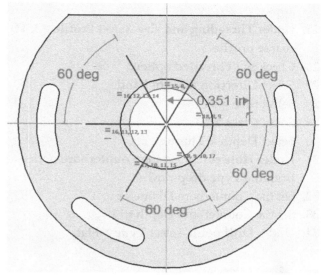

the other lines, as shown.

8. Click the **Constrain Equal** ≡ icon on the **Sketcher constraints** toolbar.
9. Make all the lines equal in length.
10. Click the **Constrain distance** ↗ icon on the **Sketcher constraints** toolbar.
11. Select anyone of the lines.
12. Type **0.351** in the **Length** box, and then click **OK**.

18. Click the **Toggle Construction geometry** icon on the **Sketcher geometries** toolbar.
19. Click the **Create Circle** ◉ icon on the **Sketcher geometries** toolbar.
20. Select the endpoint of anyone of the lines.
21. Move the pointer outward and click to create the circle.
22. Likewise, create circles coinciding with the endpoints of the remaining lines, as shown.

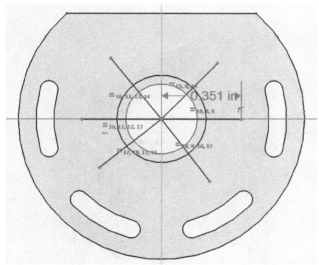

13. Click the **Constrain angle** ◁ icon on the **Sketcher constraints** toolbar.
14. Select the horizontal line.
15. Select the inclined line next to it.
16. Type **60** in the **Angle** box and click **OK**.
17. Likewise, create the angle constraints between

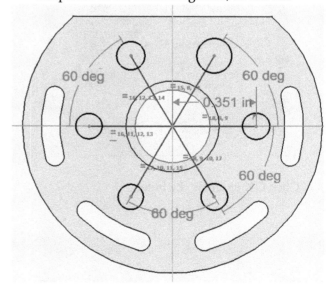

23. Click the **Close** button on the **Combo View** panel.
24. Click the **Hole** icon on the **Part Design**

Modeling toolbar; the **Hole Parameters** section appears.

25. Under **Threading and size**, select **Profile > UTS coarse profile**.
26. Check the **Threaded** option.
27. Select **Direction > Right hand**.
28. Select **Size > #1**.
29. Select **Class > 2B**.
30. Select **Depth > Through all**.
31. Under **Hole cut**, select the **Counterbore** option from the **Type** drop-down.
32. Set the Counterbore **Diameter** to 3.
33. Set the Counterbore **Depth** to 1.
34. Under **Drill point**, select **Type > Flat**.

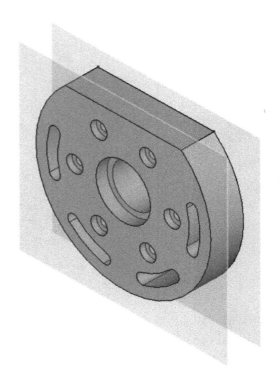

35. Click **OK** to create the hole.

Creating Chamfers

1. Select the circular edge of the counterbore hole.

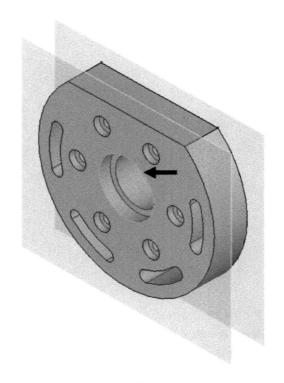

2. Click the **Chamfer** icon on the **Part Design Modeling** toolbar.

3. Enter 0.039 in the **Size** box.

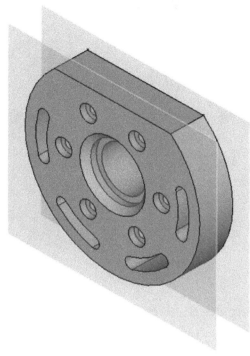

4. Click **OK** to create the chamfer.
5. Save the model and close it.

TUTORIAL 2

In this tutorial, you will create the model shown in figure.

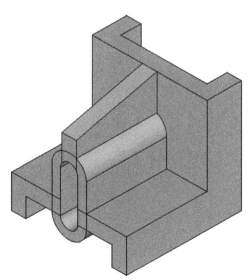

Creating the first feature

1. Open a new FreeCAD file.

2. Select the **Part Design** from the **Workbenches** drop-down.

3. Click the **Create sketch** icon on the **Part Design Helper** toolbar, and then select the YZ Plane.

4. Click **OK** to start the sketch.

5. Draw the sketch using the **Polyline** tool, as shown.

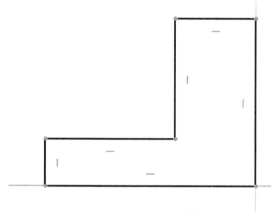

6. Click the **Constrain Equal** icon on the **Sketcher constraints** toolbar.

7. Select the vertical and horizontal lines, as shown.

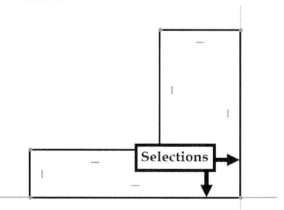

8. Select the two lines to make them equal in length.

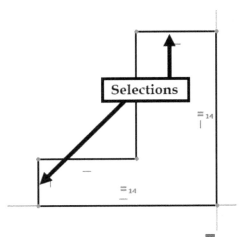

9. Click the **Constraint Vertical** $\mathbf{I}$ icon on the **Sketcher constraints** toolbar.
10. Add the constraints to the vertical lines, as shown.

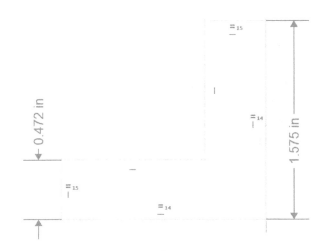

11. Click **Leave Sketch** on the **Part Design Helper** toolbar.
12. Click the **Pad** icon on the **Part Design Modeling** toolbar.
13. Select the **Symmetric plane** option from the **Pad parameters** section.
14. Set the **Length** to 1.575.
15. Click **OK** to create the first feature.

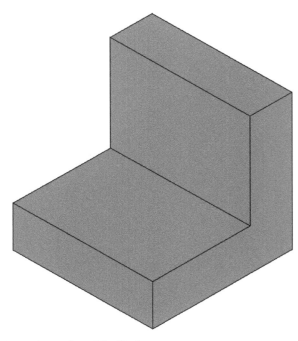

Creating the Shell feature

You can create a shell feature by removing a face of the model and applying thickness to other faces.

1. Press and hold the Ctrl key.
2. Select the top face and the back face of the model.

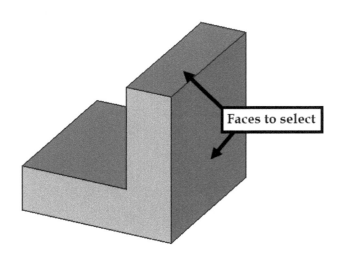

3. Click the **Thickness** icon on the **Part Design Modeling** toolbar.
4. Set **Thickness** to 0.197.
5. Select **Join type > Intersection**.
6. Check the **Make thickness inward** option.

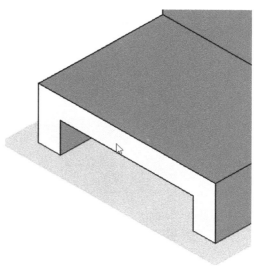

Now, you need to select the more faces.

7. Click the **Add face** button.
8. Select the front face.
9. Click the **Add face** button and the bottom face of the model.

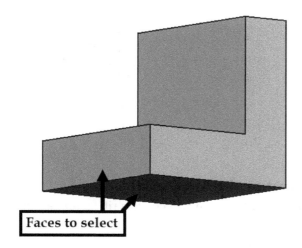

Faces to select

10. Click **OK** to shell the model.

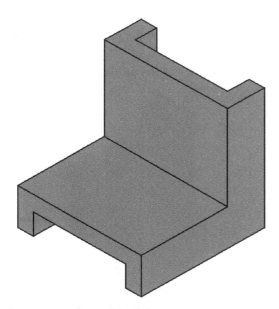

Creating the Third feature

1. Click the **Create a datum plane** ◇ icon on the **Part Design Modeling** toolbar.
2. Select the front face of the model.

3. Click **OK**.
4. Click the **Create Sketch** icon on the **Part Design Helper** toolbar.
5. Click the **External geometry** 🚰 icon on the **Sketcher geometries** toolbar.
6. Select the edges of the model, as shown.
7. Click **Create Slot** 〇 on the **Sketcher geometries** toolbar.
8. Draw a slot by selecting the first and second points.

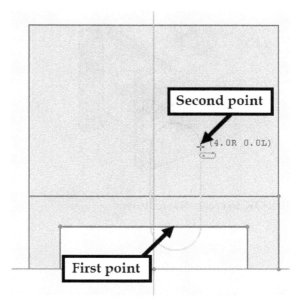

Second point

(4.0R 0.0L)

First point

9. Add constraints to the slot.

57

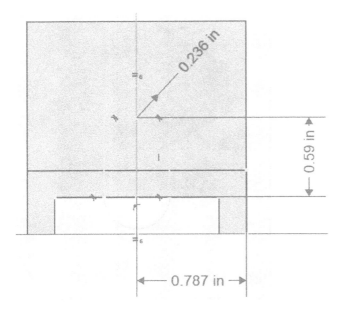

10. Click **Leave Sketch** on the **Part Design Helper** toolbar.
11. Click the **Pad** icon on the **Part Design Modeling** toolbar.
12. Select the **Up to face** option from the **Type** drop-down.
13. Select the back face of the model.

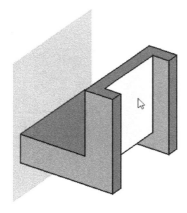

14. Click **OK** to create the feature.

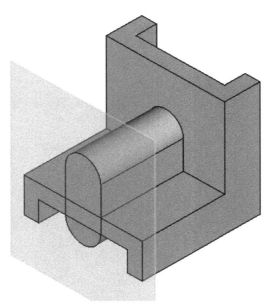

Creating the Rib Feature

In this section, you will create a rib feature at the middle of the model. To do this, you must create an offset plane.

1. To create an offset plane, click the **Create a datum plane** icon on the **Part Design Helper** toolbar.
2. Select the right face of the model.

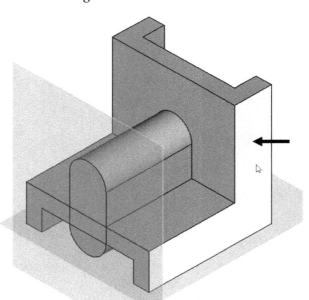

3. Type **0.7875** in the **Z** box available in the **Attachment** section.

4. Check the **Flip sides** option.

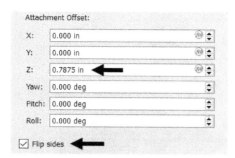

5. Click **OK** to create the plane.

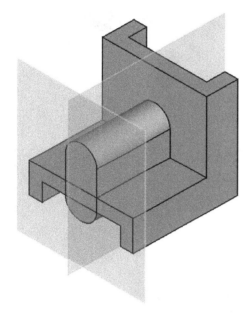

6. Click the **Create Sketch** icon on the **Part Design Helper** toolbar.
7. Select the newly created plane.
8. Click **OK**.
9. On the **View** toolbar, set the **Draw Style** to **Wireframe** .
10. Click the **External Geometry** icon on the **Sketcher geometries** toolbar.
11. Select the model edges, as shown.

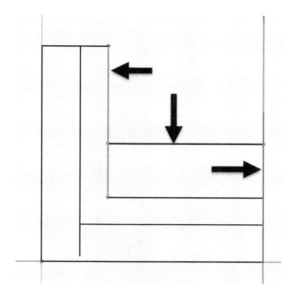

12. Draw the sketch, as shown below.

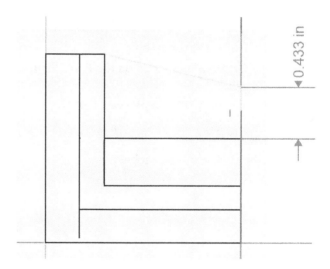

13. Click **Close** on the **Combo View** panel.
14. On the **View** toolbar, set the **Draw Style** to **Flat Lines** .
15. Click the **Pad** icon on the **Part Design Modeling** toolbar.
16. Check the **Symmetric to plane** option.
17. Type **0.197** in the **Length** box.
18. Click **OK** to create the rib feature.

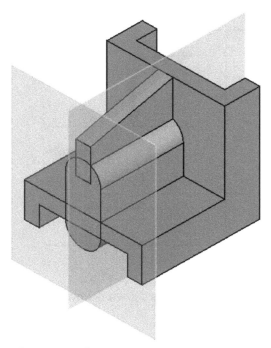

Creating a Pocket Feature

1. Click the **Create Sketch** icon on the **Part Design Helper** toolbar.
2. Select the datum plane displayed on the front face.
3. Click **OK**.
4. Create the sketch, as shown below.

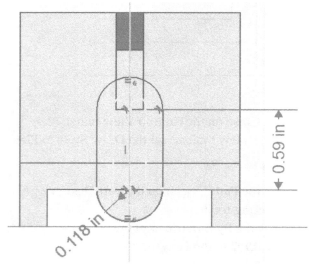

5. Click **Close** on the **Combo View** panel.
6. Click the **Pocket** icon on the **Part Design Modeling** toolbar.
7. Select the **Through All** option from the **Type** drop-down.
8. Click **OK** to create the pocket feature.

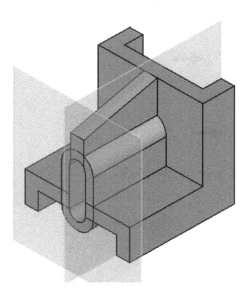

19. Save the model and close it.

TUTORIAL 3

In this tutorial, you will create a helical spring.

Creating the Profile

1. Open a new FreeCAD file.
2. Select **Part Design** from the **Workbenches** drop-down.
3. Select the XZ Plane from the **Combo View** panel.
4. Click **OK**.

5. Create a circle, as shown.

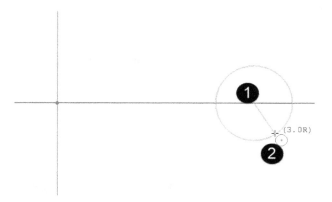

6. Add constraints to it, as shown.

7. Click **Close** on the **Combo View** panel.

Creating the Helix

1. Select **Part** from the **Workbenches** drop-down.
2. Click the **Create primitives** icon on the **Solids** toolbar.
3. In the **Geometric Primitives** dialog, select Helix from the drop-down.
4. Type **0.59** in the **Pitch** box.
5. Type **4.72** in the **Height** box.
6. Type **0.7875** in the **Radius** box.
7. Select **Right-handed** from the **Coordinate system** drop-down.

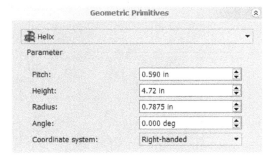

8. Expand the **Location** section.
9. Type **0** in the **X**, Y, and **Z** boxes.
10. Select **Z** from the **Direction** drop-down.
11. Click the **Create** button.
12. On the **View** toolbar, click the **Axonometric** icon.

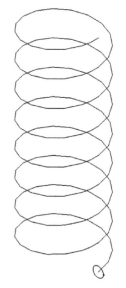

13. Click the **Close** button.

Creating the Sweep

1. Click the **Sweep** icon on the **Part tools** toolbar.
2. Select the **Sketch** from the **Available profiles** section.
3. Click the **Add** button.
4. Check the **Create solid** option.
5. Click the **Sweep Path** button.
6. Press and hold the Ctrl key and select all the elements of the helix.
7. Click **Done**.
8. Click **OK** to create the sweep.

9. Save the model and close the file.

TUTORIAL 4

In this tutorial, you create a shampoo bottle using the **Loft**, **Pad**, and **Sweep** tools.

Creating the Loft feature

To create a loft feature, you need to create sections.

1. Start a new FreeCAD file.
2. Select the **Part Design** option from the **Workbenches** drop-down.
3. Click the **Create Sketcher** icon on the **Part**

Design Helper toolbar.

4. Select the XY Plane and click **OK**.
5. Click the **Create ellipse by center** icon on the **Sketcher geometries** toolbar.
6. Draw the ellipse by selecting the points, as shown.

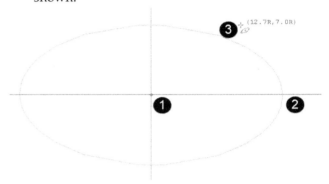

7. Apply the constraints to the ellipse.

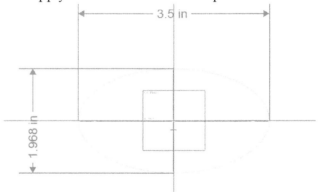

8. Click the **Close** button on the **Combo View** panel.
9. Click the **Create a datum plane** ◇ icon on the **Part Design Helper** toolbar.
10. Click on the ellipse.
11. Type 1.5 in the **Z** box.
12. Click **OK**.

13. Click the **Create a datum plane** 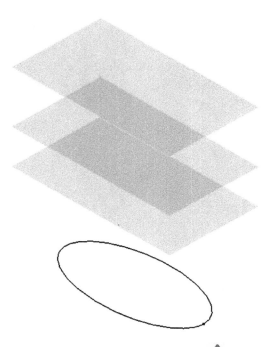 icon on the **Part Design Helper** toolbar.
14. Click on the newly created plane.
15. Type 1.5 in the **Z** box.
16. Click **OK.**

17. Click the **Create a datum plane** icon on the **Part Design Helper** toolbar.
18. Click on the newly created plane.
19. Type 1.2 in the **Z** box.
20. Click **OK.**

21. Click the **Create a datum plane** icon on the **Part Design Helper** toolbar.
22. Click on the newly created plane.
23. Type 1.2 in the **Z** box.
24. Click **OK.**

25. Click the **Create a datum plane**  icon on the **Part Design Helper** toolbar.
26. Click on the newly created plane.
27. Type 1.5 in the **Z** box.
28. Click **OK**.

29. Click the **Create a datum plane** icon on the **Part Design Helper** toolbar.
30. Click on the newly created plane.
31. Type 1.2 in the **Z** box.
32. Click **OK**

33. Click the **Create Sketch** icon on the **Part Design Modeling** toolbar.
34. Select the first datum plane, and then click **OK**.

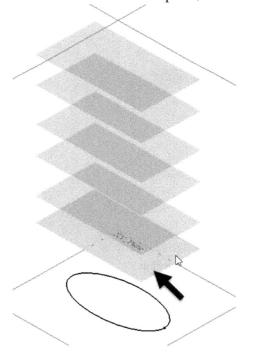

35. Create an ellipse, as shown.

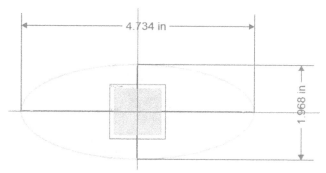

36. Click the **Close** button on **Combo View** panel.
37. Click the **Create Sketch** 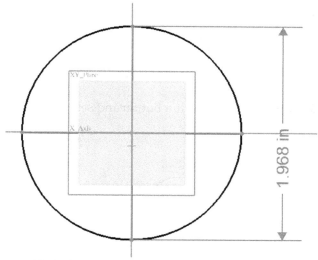 icon on the **Part Design Helper** toolbar.
38. Select the second datum plane, and then click **OK**.
39. Create an ellipse, as shown.

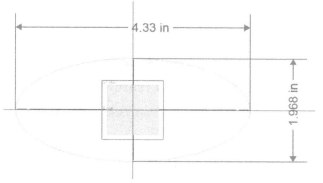

40. Click the **Close** button on the **Combo View** panel.
41. Start a sketch on the third datum plane.
42. Create an ellipse, as shown.

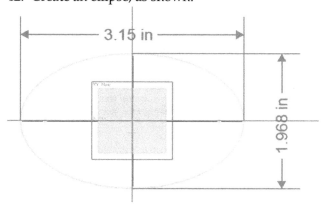

43. Click the **Close** button on the **Combo View** panel.
44. Start a sketch on the fourth datum plane, and then create an ellipse on it.

45. Click the **Close** button on the **Combo View** panel.
46. Start a new sketch on the fifth plane.
47. Create an ellipse, as shown.

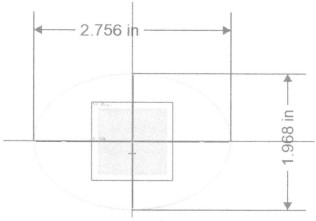

48. Click the **Close** button on the **Combo View** panel.
49. Start a new sketch on the sixth plane.
50. Create a circle, as shown.

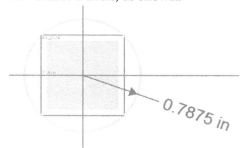

51. Click the **Close** button on the **Combo View** panel.

52. Right click to deselect the last sketch.
53. Click the **Additive Loft** icon on the **Part Design Modeling** toolbar.
54. Select the ellipse located at the bottom.
55. Click **OK**.
56. Click the **Add section** button and select the next ellipse.

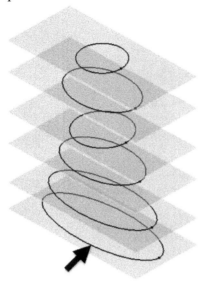

57. Likewise, select the remaining sketches in the order, as shown.

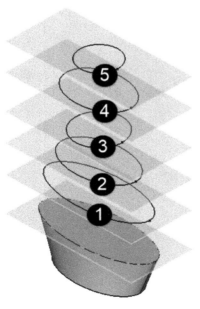

58. Click **OK** to create the loft feature.

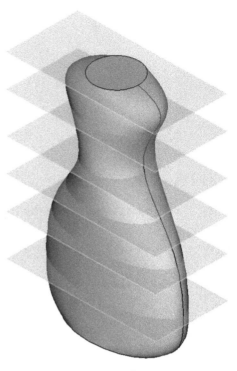

Creating the Extruded feature

1. Click the **Pad** icon **Part Design Modeling** toolbar.
2. Check the **Allow used features** option from the **Select Features** dialog.
3. Select the last sketch from the sketch list.

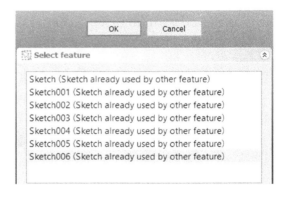

4. Click **OK**.
5. Type **1** in the **Length** box and click **OK**.

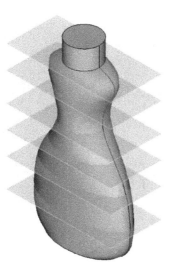

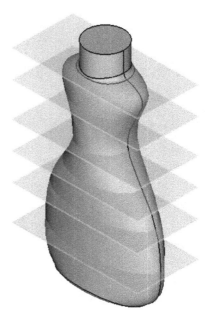

Creating Fillets

1. Press and hold the Ctrl key.
2. Click on the bottom and top edges of the swept feature.

5. Click **OK**.

Shelling the Model

1. Select the top face of the cylindrical feature.

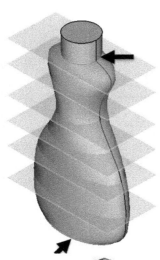

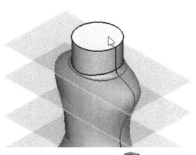

3. Click the **Fillet** icon on the **Part Design Modeling** toolbar.
4. Set **Radius** to 0.2.

2. Click the **Thickness** icon on the **Part Design Modeling** toolbar.
3. Set **Thickness** to 0.03.
4. Click **OK** to create the shell.

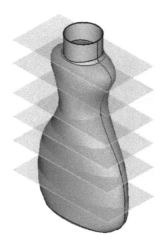

Adding Threads

1. Click the **Create Sketch** icon on the **Part Design Helper** toolbar.
2. Select the XZ Plane, and then click **OK**.
3. Click **Draw Style > Wireframe** on the **View** toolbar.
4. Click the **Toggle Construction geometry** icon on the **Sketcher geometries** toolbar.
5. Click the **Create Polyline** icon on the **Sketcher geometries** toolbar.
6. Select the origin point of the sketch, move the cursor vertically upward and click to create a vertical construction line.

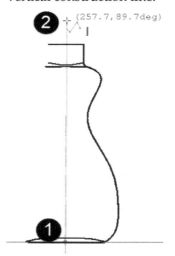

7. Right click to end the chain.
8. Create a horizontal construction line, as shown.

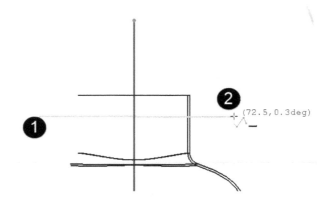

9. Click the **Toggle construction geometry** icon on the **Sketcher geometries** toolbar.
10. Create a closed profile, as shown.

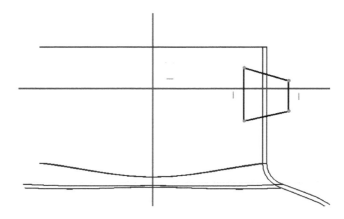

11. Press Esc to deactivate the **Create Polyline** tool.
12. Draw the thread profile.

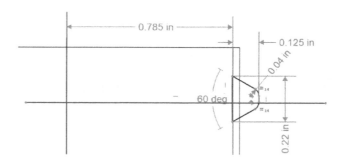

13. Click the **Constrain symmetrical** icon on the **Sketcher constraints** toolbar.
14. Select the right vertical line of the profile.
15. Select the right endpoint of the construction line.

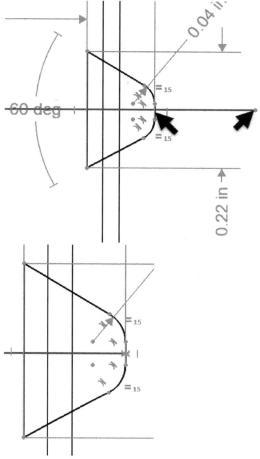

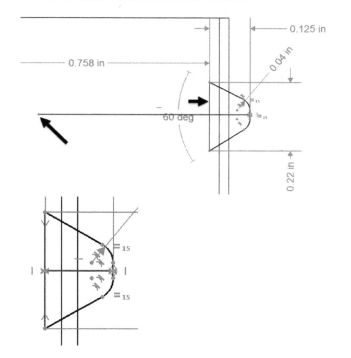

16. Select the left vertical line and the left endpoint of the horizontal construction line.

17. Create a **Vertical distance** constraint, as shown.

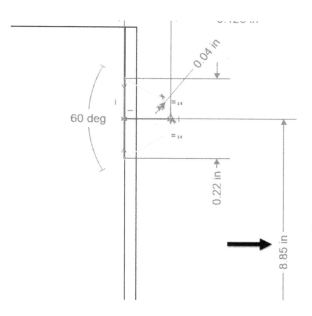

18. Click the **Close** button on the **Combo View** panel.
19. Select **Draw Style > Flat Lines** from the View toolbar.
20. Select **Part** from the **Workbenches** drop-down.
21. Click the **Create primitives** icon on the **Solids** toolbar.
22. In the **Geometric Primitives** dialog, select Helix from the drop-down.
23. Type **0.27** in the **Pitch** box.
24. Type **0.55** in the **Height** box.
25. Type **0.785** in the **Radius** box.
26. Type **90** in the **Angle** box.
27. Select **Right-handed** from the **Coordinate system** drop-down.

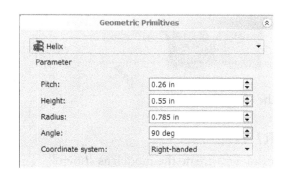

28. Expand the **Location** section.
29. Type **0,0, 8.35** in the **X, Y,** and **Z** boxes, respectively.
30. Select **Z** from the **Direction** drop-down.

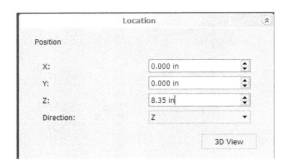

31. Click the **Create** button.
32. Click the **Close** button.
33. On the **View** toolbar, click the **Axonometric** icon.
34. Click the **Sweep** icon on the **Part tools** toolbar.
35. Select the last sketch from the **Available profiles** section.
36. Click the **Add** button.
37. Check the **Create solid** and **Frenet** options.
38. Click the **Sweep Path** button.
39. Press and hold the Ctrl key and select all the elements of the helix.
40. Click **Done**.
41. Click **OK** to create the sweep.

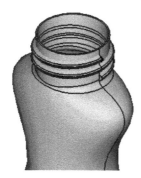

42. Save the model.

TUTORIAL 5

In this tutorial, you create the model, as shown.

Creating the First feature

1. Open a new FreeCAD file.
2. Select **Part Design** from the **Workbenches** drop-down.
3. Click the **Create Sketch** icon on the **Part Design Helper** toolbar.
4. Select the YZ Plane and click **OK**.
5. Create the sketch, as shown.

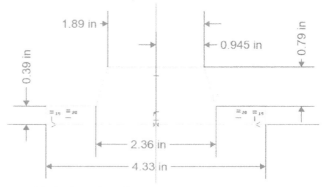

6. Click the **Close** button on the **Combo View** panel.
7. Click the **Pad** icon on the **Part Design Modeling** toolbar.
8. Type **1.575** in the **Length** box.
9. Check the **Symmetric to plane** option.
10. Click **OK**.

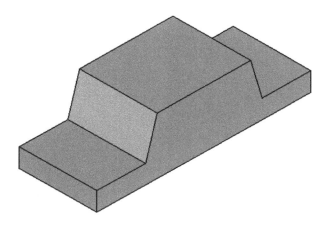

Creating the Second feature

1. Click the **Create Sketch** icon on the **Part Design Helper** toolbar.
2. Select the XY Plane and click **OK**.
3. Select **Draw Style > Wireframe** from the **View** toolbar.
4. Click the **Create Polyline** icon on the **Sketcher geometries** toolbar.
5. Create a horizontal line, as shown.

6. Press the M key thrice; the Arc tool is activated.
7. Move the pointer upward and click to create a tangent arc.

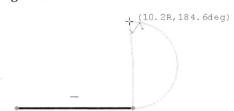

8. Move the pointer toward left and click to create a line.

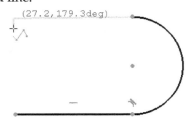

9. Select the start point of the polyline.

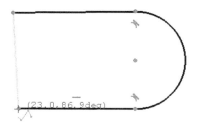

10. Apply the **Horizontal** and **Vertical** constraints to the lines, as shown.

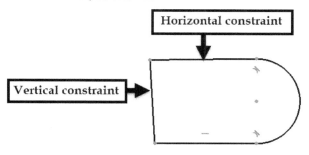

11. Click the **Constrain point onto object** icon on the **Sketcher constraints** toolbar.
12. Select the centerpoint of the arc.
13. Select the horizontal axis of the sketch.

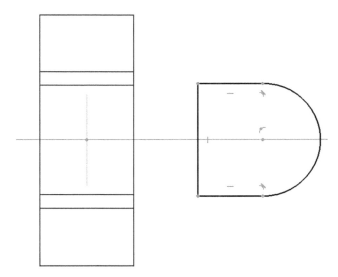

14. Click the **External geometry** icon on the **Sketcher geometries** toolbar.
15. Select the right vertical edge of the model, as shown.

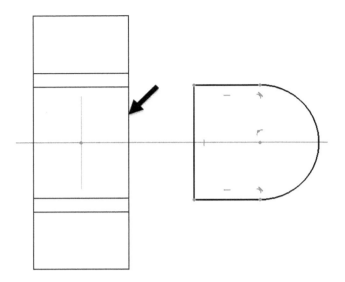

16. Click the **Constrain point onto object** icon on the **Sketcher constraints** toolbar.
17. Select the endpoint of the vertical line of the sketch, as shown.
18. Select the external geometry element.

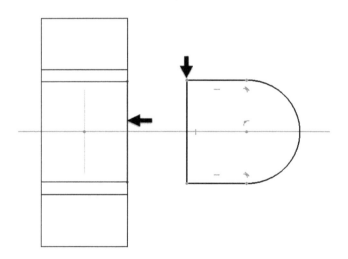

19. Create a circle by selecting the center point of the arc, as shown.

20. Add remaining constraints to the sketch, as shown.

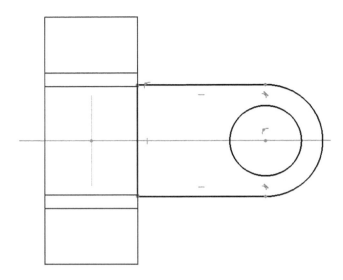

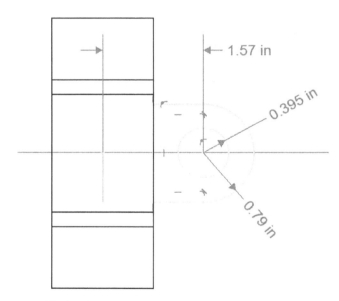

21. Click **Close** on the **Combo View** panel.
22. Click the **Pad** icon on the **Part Design Modeling** toolbar.
23. Type **0.39** in the **Length** box.
24. Click **OK**.
25. Select **Draw Style > Flat Lines** from the **View** toolbar.

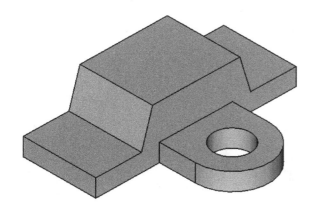

Creating the Mirrored feature

1. Click the **Mirrored** icon on the **Part Design Modeling** toolbar.
2. Select the **Pad001** feature from the **Select feature list**.
3. Click **OK**.
4. Select **Base YZ plane** from the **Plane** drop-down.
5. Click **OK**.

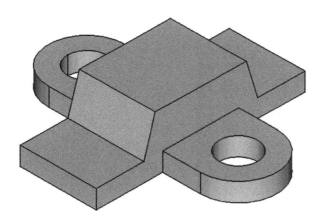

Creating Clones

1. Select the **Body** from the **Model** tab of the **Combo View** panel.

2. Click the **Create a clone** icon on the **Part**

Design Helper toolbar; the clone of the entire body is created.

3. Select the **Clone** feature from the **Model** tab of the **Combo View** panel.

4. Click the button next to the **Placement** box.

5. In the **Placement** section, select Y from the **Axis** box under the **Rotation** section.
6. Type **180** in the **Angle** box and click **OK**.

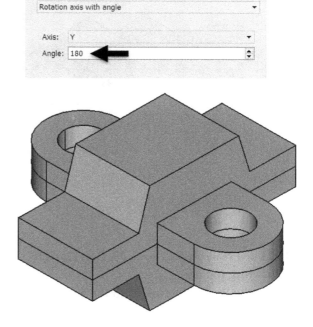

Creating the Pocket Feature

1. Click the **Create Sketch** icon on the **Part Design Helper** toolbar.
2. Select the XY Plane and click **OK**.
3. Change the **Draw Style** to **Wireframe**.
4. Create a rectangle and add constraints to it, as shown.

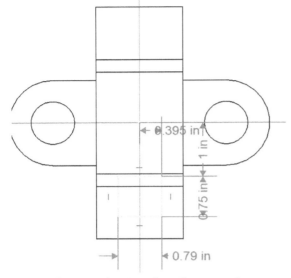

5. Click **Close** on the **Combo View** panel.
6. Click the **Pocket** icon on the **Part Design Modeling** toolbar.
7. Select **Type > Through All**.
8. Check the **Reversed** option.
9. Click **OK**.

6. Click the **Mirrored** icon on the **Part Design Modeling** toolbar.
7. Select the **Pocket** feature from the **Select feature** list, and then click **OK**.
8. Select **Base XZ plane** from the **Plane** drop-down.
9. Click **OK** to mirror the pocket feature.

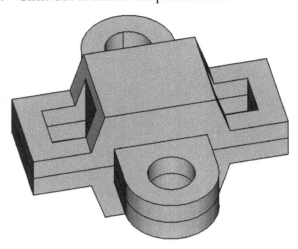

10. Save and close the file.

TUTORIAL 6

In this tutorial, you construct a patterned cylindrical shell.

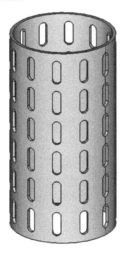

Constructing a cylindrical shell

1. Start a new FreeCAD file.
2. Select Part Design from the Workbenches drop-down.
3. Create a sketch on the XY plane.

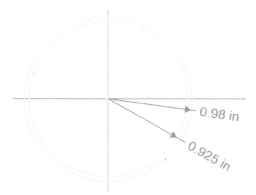

4. Extrude the sketch up to 3.93 depth.

Adding a Slot

1. Activate the **Create Sketch** tool.
2. Select the **XZ Plane** and click **OK**.
3. Set the **Draw Style** to **Wireframe**.
4. Click the **Create Slot** icon on the **Sketcher geometries** toolbar.
5. Click on the vertical axis of the sketch to define the first point of the slot.
6. Move the pointer up and click to define the second point.

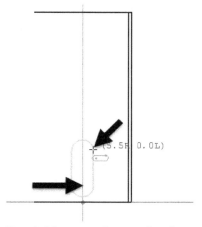

7. Add constraints to the slot.

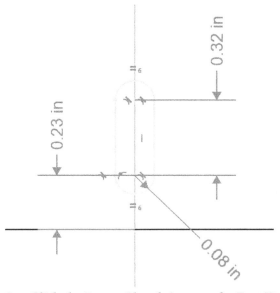

8. Click the **Leave Sketch** icon on the **Part Design Helper** toolbar.
9. Set the **Draw Style** to **Flat Lines**.
10. Click the **Pocket** icon on the **Part Design Modeling** toolbar.
11. On the **Pocker Parameters** section, select **Type > Through All**.
12. Check the **Reversed** option.
13. Click **OK**.

Constructing the Linear pattern

1. Click the **Linear Pattern** icon on the **Part Design Modeling** toolbar.
2. Select the **Pocket** feature from the **Select feature** list, and then click **OK**.
3. On the **Linear Patter parameters** section, select **Base Z axis** from the **Direction** drop-down.
4. Type **3.145** in the **Length** box.
5. Type **6** in the **Occurrences** box.

6. Click **OK**.

Constructing the Circular pattern using the MultiTransform tool

1. In the **Model** tab of the **Combo View** panel, right click on the **LinearPattern** feature, and then select **Create MultiTransform**.

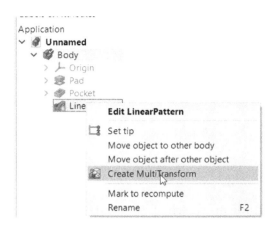

2. In the **MultiTransform parameters** section, right click in the **Transformations** section, and then select **Add polar pattern**.
3. Select **Base Z axis** from the **Axis** drop-down.
4. Type **360** in the **Angle** box.
5. Type-in **12** in the **Occurrences** box.
6. Click **OK** to make the circular pattern.

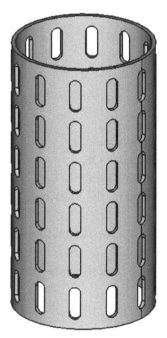

11. Save and close the model.

TUTORIAL 7

In this tutorial, you construct a pulley wheel using the **Revolution** and **Groove** tools.

1. Open a new FreeCAD file.
2. Select **Part Design** from the **Workbenches** drop-down.
3. Click the **Create Sketch** icon on the **Part Design Helper** toolbar.
4. Select the YZ plane and click **OK**.
5. Click the **Create Polyline** icon on the **Sketcher geometries** toolbar.
6. Create a closed sketch, as shown.

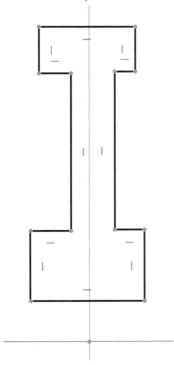

7. Click the **Constrain Equal** icon on the **Sketcher constraints** toolbar.
8. Make the entities of the sketch equal in length, as shown.

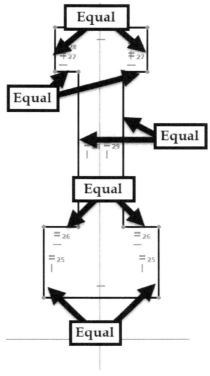

9. Create remaining constraints, as shown.

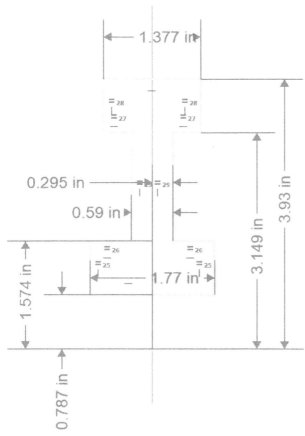

10. Click the **Close** button on the **Combo View** panel.
11. Click the **Revolution**  icon on the **Part Design Modeling** toolbar; the sketch is selected automatically.
12. Select the Y axis from the graphics window.
13. Type 360 in the **Angle** box.
14. Click **OK** to construct the revolved feature.

Constructing the Groove feature

1. Click the **Create Sketch** icon on the **Part Design Helper** toolbar.
2. Select the YZ Plane and click **OK**.
3. Set the **Draw Style** to **Wireframe**.
4. Create the sketch, as shown.

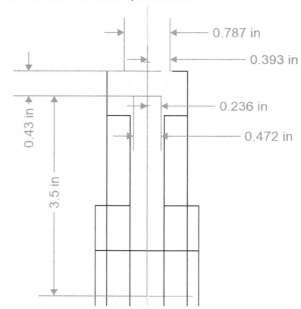

5. Click the **Close** button on the **Combo View** panel.
6. Set the **Draw Style** to **Flat Lines**.
7. Click the **Groove** icon on the **Part Design geometries** toolbar.
8. Select the Y-axis from the graphics window.

78

9. Type 360 in the **Angle** box.
10. Click **OK**.

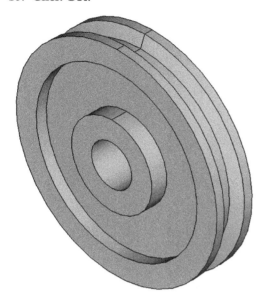

11. Save and close the model.

Additional Modeling Tools

Chapter 6: Creating Drawings

In this chapter, you will generate 2D drawings of the parts. You will learn to:

- Insert standard views of a part model
- Add dimensions

TUTORIAL 1

In this tutorial, you will create the drawing of Tutorial 7 file created in the fourth chapter.

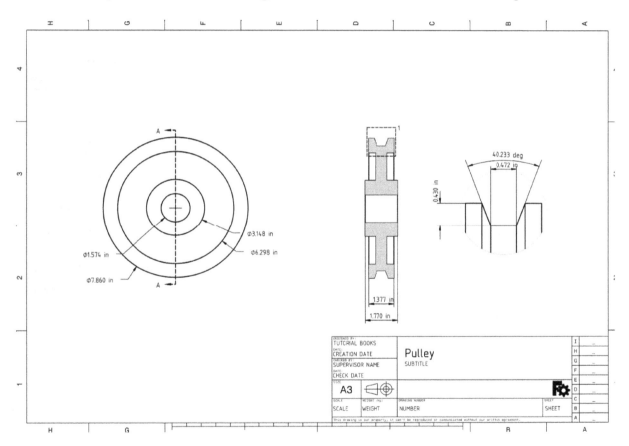

Starting a New Drawing File

1. Click **File > Open** on the Menu bar.

2. Go to the location of the Tutorial 7 file of Chapter 4.

3. Select the Tutorial 7 file and click the **Open** button.

4. Select **TechDraw** from the **Workbenches** dropdown.

5. Click the **Insert new drawing page from template** icon on the **TechDraw Pages** toolbar.
6. Select the **A3_LandscapeTD** template.
7. Click **Open**.
8. In the **Combo View** panel, select the **Page** from the **Model** tab.
9. In the **Properties** section of the **Combo View** panel, select **Projection Type > Third Angle**.

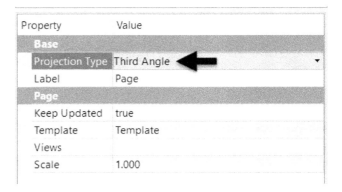

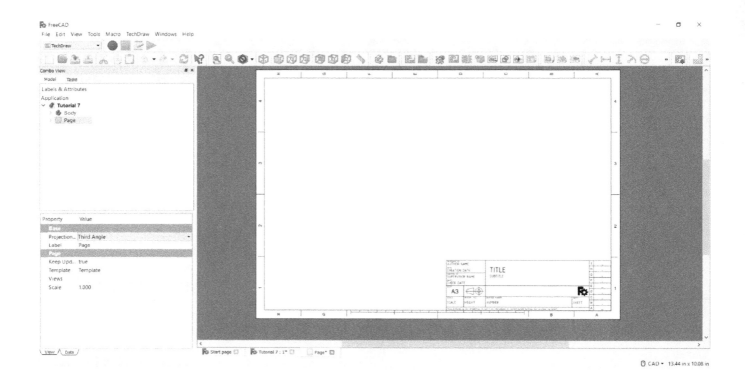

Generating the Base View

1. Select **Body** from the **Model** tab of the **Combo View** panel.
2. To generate the base view, click the **Insert Projection Group** 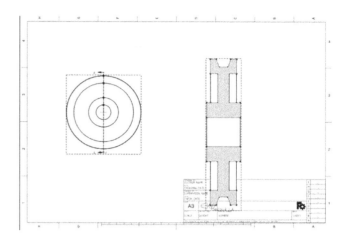 icon on the **TechDraw Views** toolbar.
3. On the **Projection Group** section, select **Scale > Custom**.
4. Type **1** and **2** in the **Scale Numerator** and **Scale Denominator** boxes located next to **Custom Scale**.
5. Click **OK** on the **Combo View** panel.
6. Click on the dotted border line of the view.
7. Press and hold the left mouse and the drag the view to left.

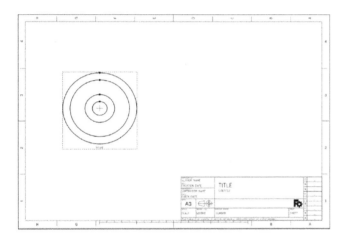

Generating the Section View

1. Select the base view.
2. Click the **Insert section view in drawing** icon on the **TechDraw Views** toolbar.
3. Click the **Looking left** icon on the **Quick Section Parameters** section.
4. Type **0** in the **X**, **Y**, and **Z** boxes, respectively. These values define the location of the section plane.
5. Click **OK** to create the section view.

6. Select the section view from the drawing or **Combo View** panel.
7. In the **Properties** section, select **Scale Type > Custom**.
8. Type 0.5 in the **Scale** field.
9. Click in the drawing sheet to update the section view.
10. Drag the section view and position it properly. Make sure that it is horizontally in-line with the base view.

Creating the Detailed View

Now, you have to create the detailed view of the groove, which is displayed, in the section view.

1. Select the section view.
2. Click the **Insert detail view in drawing** icon on the **TechDraw Views** toolbar.

The detailed view is generated, as shown. Now, you need to specify the portion of the section view to be displayed in the detailed view.

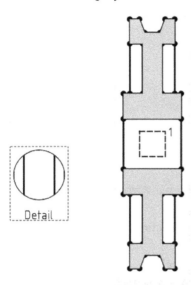

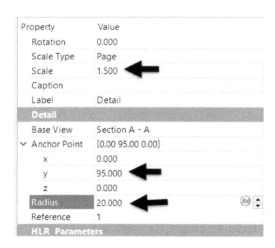

3. Select the detailed view.
4. In the **Properties** section, expand the **Anchor point** drop-down in the **Detail** section.
5. Type **95** in the **y** box.
6. Type **30** in the **Radius** box.
7. Type **1.5** in the **Scale** box.

Property	Value
Rotation	0.000
Scale Type	Page
Scale	1.500
Caption	
Label	Detail
Detail	
Base View	Section A - A
⌄ Anchor Point	[0.00 95.00 0.00]
x	0.000
y	95.000
z	0.000
Radius	20.000
Reference	1
HLR Parameters	

8. Drag the detailed to the right side on the drawing sheet.

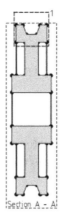

Section A - A

Adding Dimensions

Now, you will add dimensions to the drawing.

1. Select the outer circular edge of the base view.
2. Click the **Diameter dimension** icon on the **TechDraw Dimensions** toolbar.
3. Click and drag the dimension outside the view.

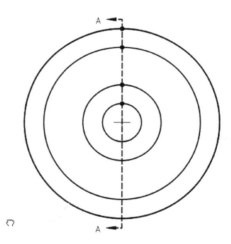

4. Likewise, add remaining dimensions to the base view, as shown.

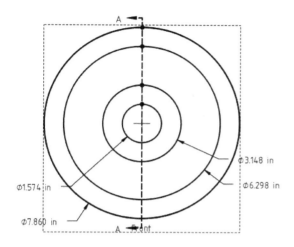

5. Press and hold the Ctrl key and select the vertices of the section view, as shown.

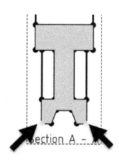

6. Click the **Horizontal-distance dimension** icon on the **TechDraw Dimensions** toolbar.

7. Press and hold the Ctrl key and select the vertices of the section view, as shown.

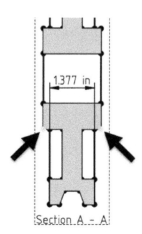

8. Click the **Horizontal-distance dimension** icon on the **TechDraw Dimensions** toolbar.

9. Drag the dimensions downward, as shown.

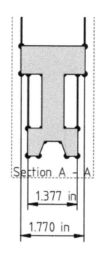

10. Press and hold the Ctrl key and select the two inclined lines in the detailed view, as shown.

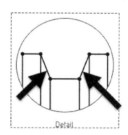

11. Click the **Angle Dimension** icon on the **TechDraw Dimensions** toolbar.

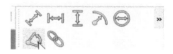

12. Press and hold Ctrl key and select the vertices of the detailed view, as shown.

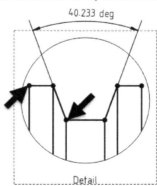

13. Click the **Vertical-distance dimension** $\underline{\overline{\text{I}}}$ icon on the **TechDraw Dimensions** toolbar.

14. Create the remaining dimension, as shown.

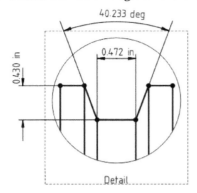

Populating the Title Block

1. Zoom in to the title block area.

2. Double-click on the green square displayed on TITLE.

3. Type **Pulley** in the **Value** box.

4. Click **OK**.

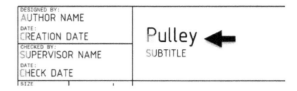

5. Likewise, add data to the remaining fields in the Title block.

6. Save and close the file.

www.ingramcontent.com/pod-product-compliance
Lightning Source LLC
Chambersburg PA
CBHW060454060326
40689CB00020B/4530